T0191893

Soil Forensics

Series editor
Henk Kars
Faculty of Earth and Life Sciences
VU University Amsterdam
Amsterdam, The Netherlands

To be a forum for all (scientific) workers in the rather fragmented field of Soil Forensics. This fragmented character is intrinsic to multidisciplinary research fields and a common platform for the exchange of knowledge and discussion is therefore heavily needed. To promote the field of Soil Forensics in academia, in forensic research institutes, legal profession/jurisdiction organisations and for the general public (science sections in newspapers). To contribute to a high scientific standard of the field. To be attractive for publishing in the series it is peer reviewed in order to be competitive with journals such as Forensic Science International.

More information about this series at http://www.springer.com/series/11807

Rosa Maria Di Maggio • Pier Matteo Barone
Editors

Geoscientists at Crime Scenes

A Companion to Forensic Geoscience

 Springer

Editors
Rosa Maria Di Maggio
Geoscienze Forensi Italia® – Forensic
  Geoscience Italy
Rome, Italy

Pier Matteo Barone
Archaeology and Classics Program
American University of Rome
Rome, Italy

Geoscienze Forensi Italia® – Forensic
  Geoscience Italy
Rome, Italy

ISSN 2214-4293   ISSN 2214-4315   (electronic)
Soil Forensics
ISBN 978-3-319-86310-8   ISBN 978-3-319-58048-7   (eBook)
DOI 10.1007/978-3-319-58048-7

Printed on acid-free paper

This Springer imprint is published by Springer Nature
The registered company is Springer International Publishing AG
The registered company address is: Gewerbestrasse 11, 6330 Cham, Switzerland

Preface

Geoscientists at Crime Scenes is specifically dedicated to the multi-disciplinary subject of forensic geology. Interestingly, and also in Italy, there exists anecdotal accounts going back some 2000 years or more regarding Roman soldiers who inspected soil on horse hoofs to determine a geographical provenance and to help locate enemy camps.

The documented and recorded applications of geology to police investigations can be traced to the middle part of the nineteenth century. At this time, in other parts of Europe, the geological analysis of minerals or man-made items derived from geological materials assisted police with their criminal investigations, including; murder, sexual crimes and fraud. This book essentially continues this theme.

Geologia forense. Introduzione alle Geoscienze Applicate alle Indagini Giudiziarie, published in 2013 inspired the publication of Geoscientists at Crime Scenes. This has been published in the English language and provides the opportunity for the global dissemination of a wealth of information on forensic geology. In its simplest form, forensic geology may be considered as the application of geology to policing and law enforcement. In the context of *Geoscientists at Crime Scenes* this also includes; an introduction and overview of some of the basic fundamental principles of geology and how these may be applied to policing and law enforcement, geological materials, forensic pedology, the use of remote sensing in forensic investigations, forensic geomorphology, environmental forensics, forensic geophysics, forensic archaeology and geotechnical aspects of forensic geology. Furthermore, also included are some of the operational and ethical issues facing forensic geologists working at crime scenes.

Although, as noted above, geology has been applied to crime for at least a few hundreds of years, and possibly back to Roman times, however in the past decade or more there has been a revived interest in the practical applications of forensic geology and in research. This has been facilitated by, for example, the International Union of Geological Sciences (IUGS), Initiative on Forensic Geology (IFG), which has been promoting and developing forensic geology around the world. As part of this, the co-authors have participated in numerous international outreach, knowledge transfer and training events with federal police forces and geologists. The authors are well-known, authoritative, internationally highly respected forensic geoscientists. They have a combined wealth of experiences in operational crime scene investigations and forensic geology guided research.

Police and other law enforcement organization are increasingly becoming more dependent on the expertise that can be provided by forensic geologists. As such, forensic geologists are therefore likely to continue in the years ahead providing an important role assisting with certain types of criminal investigations. It is therefore expected this book will be of value to forensic geologists (geoscientists), forensic scientists, environmental scientists, geophysicists, geomorphologists, police officers, law enforcement officers and students. Accordingly, *Geoscientist at Crime Scenes* provides information and a valuable collection of cases that are relevant for the professional and practicable applications of geology to policing and law enforcement. This book should be in the collection of every forensic geo-scientist throughout the world.

International Union of Geological Sciences (IUGS)   Laurance Donnelly
Initiative on Forensic Geology (IFG) & Arup
Manchester, UK

Contents

Chapter 1
Introduction to Forensic Geoscience

Pier Matteo Barone and Rosa Maria Di Maggio

Abstract Forensic geoscience encompasses many branches of the Earth Sciences whose applications may provide a suitable environmental interpretation of the surroundings, in the context of numerous types of offences, both criminal and civil. A historical overview, the forensic applications of the different branches of geoscience, and the geo-scientific approach to crime scene are treated in this chapter with the purpose to introducing the reader to the main topics of the book.

Keywords Forensic geosciences • Crime scene investigation • Historical overview • Geoforensic approach

If forensic science is the application of technical and scientific methodologies applied to traditional categories of judicial investigations, in connection with the investigation of a crime or a social behavior, within them a special role is held by forensic geoscience, or the application of different disciplines of Earth Sciences in judicial contexts.

Forensic geoscience encompasses some branches of the earth sciences such as geology, mineralogy, geochemistry, geophysics, remote sensing, soil science and archaeology, whose applications may provide a suitable or appropriate environmental interpretation of the surroundings, in the context of numerous types of offences, both criminal and civil.

The environmental context can become part of the crime narrative in diverse situations: it can be the place where to hide a body or precious objects, or it may be a witness to a crime. Given the nature of many crime narratives carried out in outdoor areas, the environment, as it is not a closed system, plays a dual role, passive and active. Passive in the sense that it may be the principal repository of clues and evidence that the both victim and/or the offender may have been present in a given

P.M. Barone (✉)
Archaeology and Classics Program, American University of Rome, Rome, Italy

Geoscienze Forensi Italia® – Forensic Geoscience Italy, Rome, Italy
e-mail: p.barone@aur.edu

R.M. Di Maggio
Geoscienze Forensi Italia® – Forensic Geoscience Italy, Rome, Italy
e-mail: dimaggio@geologiaforense.com

© Springer International Publishing AG 2017
R.M. Di Maggio, P.M. Barone (eds.), *Geoscientists at Crime Scenes*,
Soil Forensics, DOI 10.1007/978-3-319-58048-7_1

location. Active since it can be the protagonist of the crime narrative, where, for example, it may have been abused in the design and construction phases of landfill or architectural structures, but also when it leaves proof marks on clothing and objects, such as the soil found on footwear.

The rigorous approach of Earth Sciences to judicial investigations and the growing needs to solve different types of crime in different contexts and environments, have allowed the scientists not only to improve the techniques at the crime scene and in the laboratory, but also to benefit from new approaches afferent to the field of physics, chemistry, biology and geo-archaeology. This development has also led to the implementation of a multidisciplinary approach to the crime scene, which also involves the use of materials that are not purely geological, such as entomology, microbiology, palynology, etc. The versatility in the application of earth sciences to judicial investigations has meant that they have been rightly renamed forensic geoscience.

But if the forensic sciences are widely and historically known, geosciences have started later to carve an important area in the vast panorama of investigative practice.

In April of 1856, on a train that was travelling along the railroad the Prussian Empire, a barrel containing silver coins was emptied and filled with sand. The discovery was made only at the train's final destination. One scholar collected samples of sand from the stations along the railway line where the train had carried out the stops and, with the aid of a microscope, was able to connect the sand present in the barrel with the one coming from one of the stations. The fact was reported in the journal *Science* and Art, thus documenting the first episode in which a discipline of geosciences provided useful clues to investigators to solve a criminal case, comparing geological materials, acquired as elements of evidence. However, in the following years, there was no further news of the application of geosciences in criminal cases.

Only in the late nineteenth century, the writer Sir Arthur Conan Doyle, author of the famous series of novels "The Sherlock Holmes' adventures", described in his stories the observation of traces of land to solve criminal cases. In particular, in his novel *A Study in Scarlet*, 1887, John Watson, speaking of Sherlock Holmes' skills, says: "*he is able to discern at first sight the different soilsfrom each other; shows me, in our walks, mud stains on his pants and shows me as their texture and their colorwill enable them to know in what part of London they were done*". In the novel *The Five Orange* Pips, 1891, there is another interesting passage in which Holmes observes the presence of soil rich in chalk on a pair of boots. These steps indicate that Conan Doyle was already fully aware of the value of the morphological and mineralogical properties of the soils in forensic applications. Despite that many of the methodologies described in the Sherlock Holmes novels are inapplicable in real world, probably Doyle's was the prototypical (or pioneering) approach of a forensic geoscientist which now represents, for example, the basis for the study of soils in forensic practice:

- The number of types of soil is almost unlimited;
- Soils can change their characteristics within short distances;
- All people can have traces of soil on their clothes, footwear or vehicles simply by coming in contact with such materials;

– Examination of soil material can help to locate where a person was when they came in contact with it.

Always in the last years of the nineteenth century, Hans Gross, an Austrian professor of criminology, with amazing foresight proposed in his book *Handbuch für Untersuchungsrichter*, published in 1893 and translated into English under the title *"Criminal Investigation"*, the application of scientific methods to support investigations in criminal cases, including earth sciences and mineralogy to study, he said, *"dustand dirt present on the shoes and clothing of a suspect"*.

In the early years of the twentieth century, Georg Popp, a German chemist, solved a number of murder cases using mineralogy. Particularly in the case of the murder of Margarethe Filbert, he was able to establish that the time sequence of soil deposits on the clothes of the suspect were compatible with the homicide narrative. This was the first time that the value of geological surveys in a court of law was recognised.

Between 1910 and 1912, Edmond Locard, a French physician and criminologist, focused his scientific studies on the solving of criminal cases concentrating his interest on how various materials, including geological ones, could come in contact with and move on various surfaces in the form of traces, postulating the famous principle of exchange that today takes his name: *"whenever two objects come in contact with one another, there is always a material transfer; although the analytical methods may not be sensitive enough to demonstrate this exchange or certain conditions can eliminate it, the material transfer does take place"*. This principle forms the basis of evidence collection and analysis of all traces considered sources of evidence, form fingerprints to soil.

Throughout the twentieth century, forensic geology laboratories were set up in various police forensic departments around the world, starting with the Federal Bureau of Investigation (FBI) in 1939. In the United States, the analysis on soils in the context of judicial investigations gained particular popularity in the 1960s and 1970s due to the work of Walter McCrone and his McCrone Research Institute. McCrone opened up the field in the use of geoscience techniques, such as optical microscopy and of other methods, for the characterization of soils and a wide variety of diverse particles useful in the forensic context.

In Italy, forensic geology is a relatively young science, in fact, although the Forensic Science Police department was founded in 1903 thanks to the coroner Salvatore Ottolenghi, only at the beginning of the eighties they began using soil samples as evidence, driven by the need to solve important kidnapping cases. Specifically, the presence of dirt on suspicious vehicles could provide meaningful information on the journeys made by the kidnappers and the dens and hiding places frequented by them.

Although the editorial interest in forensic geology began to develop in the 1960s, the first text devoted to this topic was written and published by Murray and Tedrow only in 1975.

Despite the unequivocal importance of geological evidence in criminal investigation, the subject has been poorly treated in the last two decades of the twentieth century.

In the early years of the new millennium, forensic geology, has regained the interest of investigators, forensic institutes and universities and has experienced a great development, especially in the Anglo-Saxon world. The collaboration between investigators and research organizations has allowed for not only the improvement of analytical procedures, but also the introduction of new methods and new applications related to different branches of the earth sciences, which have shown significant potential in providing convincing case evidence.

These aspects have allowed the term 'forensic geoscience' to be coined, playing an increasingly important judicial role so that the strong professionalism of experts is increasingly in demand in the court rooms. At the same time, there has been a growth in the literature on this subject and the organization of several international meetings has allowed fruitful comparisons between experts from around the world, confirming the rapid development of the geosciences applied to many types of crimes.

Finally, various scientific and scientific-bar associations, both national and international, have created sections and working groups dedicated to forensic geoscience. For example, as part of the International Union of Geological Science (IUGS), in 2011 the Initiative on Forensic Geology was established (IFG), a working group that brought together many international experts from academia, forensic institutes and professionals, which aims to provide data and information about the geosciences applied to criminal, civil and environmental investigations, as well as to organizing conferences and training both for law enforcement agencies and for the scientific community. With the same purpose, and extensive national and international collaboration of leading experts, also the establishment of Forensic Geoscience – Italia® team[1] emerges as a unique force in the Italian forensic landscape, offering a wide range of services in support of geoscientific analysis for investigative and procedural frameworks.

1.1   The Applications of Forensic Geoscience

Pedology, geochemistry, geophysics, remote sensing, and geo-archaeology are among the disciplines of geoscience that, worldwide, have proven to be useful and effective investigative tools in the forensic field.

Pedology (forensic soil science) researches, studies and analyzes soiltraces connected to a crime. The soil can provide important information in the context of a criminal investigation, because it can be transferred between surfaces by contact and deposition. The numerous and different elements present in the soils and the extreme variability of their proportions establish that the typology of existing soils is infinite, with the obvious impossibility of finding identical samples. Soil analysis, therefore, are aimed at determining the level of comparability between the soil samples or of their common origin from the same micro-environment geological or

---

[1] www.geoscienzeforensiitalia.com

pedological. The careful analyzes of soils allows the characterization of the samples to the point that they can even distinguish whether they were collected a few meters away. For such evidence, the use of soil analysis is excellent as a support to judicial investigations.

Geochemistry has multiple applications in forensic contexts. The application of advanced techniques allows for the characterization of solid geological materials, such as soil, rocks, minerals, gems, and fluids, such as water and gas. This method, therefore, not only determines the composition of the details of materials, but allows, through the processing of obtained data, to understand their origin and evolution. Geochemical analysis, if properly applied, can be extremely useful in all those cases where it is necessary to verify and assess the environmental pollution in different matrices such as air, water and soil, by contaminants and hazardous substances. These analyzes make it possible to recognize the origin, the concentration and the possible interaction between different contaminants, providing factual evidence in the context of environmental surveys, waste disposal and safety in general. Finally, in forensictaphonomy – in the case of discovery of buried corpses – the application of geochemical methods is very effective for determining the amount of phosphorus, organic carbon and other substances in the soil, to determine the time of burial. Indeed, a body which decomposes releases certain substances to the soil (leachate), whose concentration varies depending on the time elapsed since the burial and the geological and soil characteristics of the site.

The analysis of air and satellite photographs is an effective tool for the relatively fast acquisition of geo-environmental and territorial information provided related to areas which may also be quite vast. It is evident the extent to which the application of this technique is extremely useful in forensics; it is particularly effective when it is difficult, sometimes impossible, to directly view the surveyed sites across the country with direct inspections, or when an immediate overview of a given territory is needed, to identify elements useful to the investigation: escape routes, natural and man-made obstacles, hiding places etc. Aerial photographs can provide useful information, especially if the area has undergone changes over time involving illegal activities against the environment and the territory, and environmental crimes, or crimes against the person and mankind, namely the concealment of bodies or the creation of mass graves.

Forensic geophysics is the localization and mapping of objects, bodies or cavities, of various types and sizes, hidden underground or under water, using specific geophysics tools, for judicial purposes. In forensic investigations a wide range of geophysical techniques are applied that have the potential to verify a contrast in physical properties between a target and the material in which it is buried. Generally, in forensic geophysics the use of electromagnetic equipment, such as the GPR (Ground Penetrating Radar) and the metal detector, is involved by means of which it is possible to obtain an approximate image of the variation of physical properties in the first few meters under the surface. Generally, abnormalities in the variations of the physical parameters can be potentially interpreted as coming from "foreign" buried materials. With these techniques it is therefore possible to identify and define precisely the place of concealment of the target in question, even to the extent of recognizing evidence of human occupation, anthropogenic disturbance of the soil or soilexcavation, both recent and after many years. The geophysical methods have the potential to swiftly and non-invasively investigate extensive areas where someone has tried to obliterate an underground clandestine burial or, in general, a forensictarget. Forensic geophysics is not only effective for the search and localization of weapons or metal drums, burials and bunkers, but is also suitable for identifying areas with high rates of chemical pollution, where the contaminants placed in the ground alter its physical properties as a function of their concentration and geometric distribution. A similar argument can be made also for decomposition liquids (leachate) released by buried corpses, easily identifiable by geophysical techniques as they alter the physical-chemical properties of the surrounding terrain.

Archeology, but especially geoarchaeology, are applicable in the forensic field when there is the need to search and locate clandestine graves and recover the remains, or where archaeological discoveries require forensic investigation in order to understand ancient and modern crime narratives.

The sequence of operations leading to the location and recovery of buried remains is divided into three main stages. A preliminary survey of the alleged place of burial to determine the location of exactly where to do the investigation, the actual excavation phase and the documentation and recovery of the remains as these are excavated. The reconnaissance and localization phases involve the interpretation of the land in order to search for a presumed burial site and they are carried out through the use of different instruments, both large (remote sensing) and small scale (field-walking) gear. For example, aerial photographs can easily provide important information about the area to be investigated (abnormal changes in vegetation cover over large areas or changes in soil morphology) and their use in the preliminary stages of a survey can be effective, leading to significant saving of time and energy. In situ the application of geophysical techniques allows a more precise delimitation of the buried target.

In a forensic context, it may be important to know how to properly remove any human remains, even if, often, in the excavationsites is very easy to find just scattered bones or teeth that have been picked up and moved by animal scavengers. The recovery of the remains by forensic geoarchaeology techniques, which combine the typical ones of the excavation with the demands of scientific and judicial inquiry, can provide information about the mode of deposition-concealment of the body and allow the discovery of clues to the investigative police.

1.2 Crime Scene Investigations

The crime scene can be defined as the place where a crime was committed and/or the places or paths related to it, that is where the crime took place. In this sense, the crime scene is not only a territorial physical space where there was a crime or a phase of it, but also a conceptual situation characterized by the behavior of the crime perpetrator and/or of his victim; a representation of events, behaviors and relationships that leave traces and information to interpret and process.

Generally, the scene of the crime is divided into zones, primary and secondary, whose distinction is not based on the order of importance, but on the sequence of the places involved in the crime narratives. The primary crime scene is the one where the crime physically took place or where a body was found, while the secondary crime scene is connected to the crime at various levels, for example the place where there is a high probability that the perpetrator completed the actions, or where the victim acted during a time preceding the crime. There is, finally, another area, called the zone of investigative interest, where the perpetrator may have transited leaving significant traces.

The scene of the crime can provide basic information to direct the investigation towards investigative leads related to the crime committed and therefore to identify the narrative and the offender. The study and analysis of the crime scene involve a process that necessarily develops with a backward path, namely that starts at the final event to try to arrive at the causes and circumstances that generated it.

The scene of the crime contains trace, physical, chemical, biological and behavioral information that need to be researched, interpreted, analyzed and processed to get, to identify and define identifiers for the author of the crime, and as such, must be preserved to optimally carry out the site technical survey.

The technical survey involves a set of activities of scientific nature, which has as its goal the preservation of the locations, the search and the preservation of elements and of the relevant traces of the crime, useful for the identification of the offender and/or the victim, as well as for the reconstruction of event narrative and for verification of the circumstances in which it was made. Therefore, the site technical survey is an act of direct investigation, through which police forces proceed in their observations of the crime scene in order to understand the narrative of the crime and collect any evidence. Considering the definition of site technical inspection and the purpose for which it is undertaken, it is clear that the closer in time it is executed close to the event increasers the possibility that the crime scene shall not be impaired and not impaired in relation to the time when the crime was committed.

The site technical inspection activities are the observation and description of locations and elements, planimetric surveys, video-photographic surveys, research and collection of trace clues and their conservation.

The observation and the subsequent description tasks are the most delicate phases of a sitesurvey because through them the state of the locations and elements are fixed or "frozen" for future reference. The observations should be as objective as possible with the possibility of recording all the details of the scene. Since the observations are performed on a subjective basis and that the human factor and that of the context can have a considerable bearing on the optimal freezing of the scene, they must be standardized by means of a method, and thus by rules. The observations must be carried out: from the general to the particular, from the outside towards the inside, from bottom to top, from right to left. During this phase cards are placed with letters or numbers to highlight clue traces, and significant findings or abnormal situations.

The description activity is subject to specific methods, suitable for the nature of the object of observation. Observed locations and things must be described in an extremely accurate manner, with thoroughness and objectivity, without reporting personal assessments.

The planimetric and video-photographic surveys complement the descriptive surveys with different methods and provide an alternative representation of the same reality. Planimetric surveys consist of drafting a scaled drawing, oriented according to the cardinal points, of the site plan and marking on it the objects or situations that are judged likely to have an association with the criminal event. Video-photographic surveys, conducted with the rules of judicial photography, make it possible to accurately reconstruct the scene of the crime in all its aspects and to highlight, also at later stages, details that may have escaped or under-valuated during the sitesurvey.

Subsequent to the descriptive, planimetric and video-photographic surveys, which necessarily also include a first phase of clue-trace searching, the forensic technicians proceed to the collection of the same and of all those objects which constitute sources of evidence. This phase can be carried out with the application of more or less rigorous methods. Whichever method is used, the collection and preservation of sources of evidence should be undertaken with extreme care to avoid possible contamination of the clue traces and by adopting special measures that differ according to the material to be filed.

The chain of custody of the evidence exhibits is a set of procedures aimed at ensuring that both the identification and the integrity of the sources of evidence are assured during all procedural stages, ranging from the collection of clue traces at the crime scene to the laboratory tests, until their final delivery to the Judicial Authority. The chain of custody consists of recording, with dedicated paper, digital and photographic documents, the sealing and opening actions on the envelopes containing the exhibits, the processing steps to which they have been subjected as well as the operators involved.

The forensic science laboratories draw up strict procedures for evidence collection of clue traces and for their conservation, for their protection from alteration and contamination with other materials or with other similar traces and for the assurance of the chain of custody. The appropriate collection of the sources of evidence and their optimal conservation, through the above procedures, guarantee the evidential and probative validity of the information obtained in the analytical phases.

1.3   The Geoscientific Approach to Crime Scene

The approach to the crime scene differs by method, quality and purpose depending on whether the operators are investigators, criminologists, forensicexperts, clue searching technical experts or criminal profiling specialists.

Even the various branches of forensic geoscience require different approaches depending on the purpose for which the technical survey is conducted and the purposes of the investigation. However, regardless of the type of the crime and the manner of intervention, the forensic geoscientist starts from a fundamental approach, common to all the geosciences, which is to study, understand, and interpret the environment and the geographical, physical and geological territory in which the actual crime took place. This is because the forensic geoscientist must be able to search for and collect adequate information from the environment, placing it in its specific criminal context and characterizing its actual or presumed narrative.

The study and analysis of the territory are carried out in two complementary stages:

- Reading the topographical, geological, geomorphological, pedological, hydro-geological and additional thematic maps, as well as the inspection of the territory with tools such as remote sensing;
- The onsite examination, which generally involves both the areas directly affected and the surrounding areas, in order to obtain the mapping and the necessary correlations of the geological, geophysical, geochemical, geotechnical and geoarchaeological characteristics of the area under consideration.

The interpretation of the territory in order to search for a presumed burial site is carried out through the use of different instruments, involving both large and small scale equipment. For example, aerial photographs can easily provide important information about the area to be investigated (abnormal changes in vegetation cover over large areas or changes in soil morphology) and their use in the preliminary stages of a survey can be effective, leading to significant saving of time and energy. On a local scale, geophysical techniques allow a more accurate positioning of the buried target.

The processing of all this information may be implemented in any geographic information system or GIS (Geographic Information System). In fact, such solutions appear to be way to link together information from different sources such as those described above, which otherwise would be impossible to correlate. The concept of GIS developed with the idea of a "container" capable of handling territorial management issues under the most varied representations, but also able to understanding and absorbing into a single inter-relational process all the automated procedures and databases associated with a forensic investigation.

The development of a spatial analysis system is manifested in different forms, because there are different ways to investigate and the interrelationships between the various protagonists, active and passive, of a crime are various. It follows that there is no a single model for territorial analysis and the tools needed to equip ourselves and achieve a proper investigation must necessarily be functional to the exploration and understanding of the interconnections between crime, territory, and the victim/perpetrator.

Each geographical context is characterized by its own internal operation scheme that depends on the history, on the cultural and social phenomena, on the geomorphological structure of the territory, on the combination of objective and subjective factors and therefore a GIS that takes into account the multidimensional nature of a territory must be able to measure magnitudes that highlight the territorial and structural diversity.

A correct and careful analysis of the local context allows for a more accurate and comprehensive resolution of the case, focusing energy and resources in certain areas of the scene under investigation. Once the site has been identified and mapped, the geo-archaeological excavation can begin, keeping in mind the concept of stratigraphy, the law of superposition and the methodology for a proper recovery.

Often, in many types of crimes the different disciplines related to the Geosciences intervene in a complementary way, with a multidisciplinary approach, to obtain essential information and/or useful sources of evidence. Consider forensictaphonomy that, in the search for buried bodies and in the study of the modes of corpse concealment, implies geophysical, geo-archaeological, soil and geochemical analytical research, to obtain space-time information about the burial and the narrative associated with the execution of the crime.

The true, and only, limiting factor in the use of geoscience in the forensic field often is the difficulty of bringing together the various skills and the difficulty of exploiting them collectively during the observation, research and acquisition stages of crime scene evidence. During such stages, in fact, it is of primary importance to optimally select those scientific techniques to be used for the search for elements and supporting information and their contextualization within the scene of the crime, the suspected criminals, and the victims. A dispersal of human resources and information caused by poor experience and cohesion of the investigation team would cause irreversible damage to the entire forensic investigation.

Chapter 2
Geological Materials on the Crime Scene

Rosa Maria Di Maggio

Abstract Geological materials, being the main constituents of the Earth's surface, regularly coming into contact with persons or are voluntarily also used in illegal activities or during the course of a crime, and they sometimes can play a primary role as evidence. This chapter covers the more significant geological materials associated with forensic investigations and it also illustrates their genetic composition and evolutionary modalities, and the way they are typically involved in criminal cases.

Keywords Minerals • Rocks • Sands • Soil • Gems • Precious metals • Fossils

The minerals, rocks, soils and all materials derived from them are in contact with humans daily and take on a primary role in many human activities ranging from industry to agriculture, from construction to leisure.

Geological materials, being the main constituents of the Earth's surface, regularly coming into contact with persons or are voluntarily also used in illegal activities or during the course of a crime, and they sometimes can play a primary role as evidence. Consider, for example, traces of soil that, during a crime, deposit to the shoes of the author of a crime; or improper use of stone materials as a blunt weapon, or exploitation of underground cavities or land features in which to hide bodies, weapons or toxic waste.

This chapter covers the more significant geological materials associated with forensic investigations and, for some of them, it also illustrates the way they are typically involved. The involvement in the forensic investigation of loose geological materials, such as sand, soil and dust, will be treated in Chap. 3. However, in order to better understand the analysis carried out on loose geological materials for forensic purposes and the potential of the obtained data, it is necessary to know their genetic composition and evolutionary modalities, which are explained in this section.

R.M. Di Maggio (✉)
Geoscienze Forensi Italia® – Forensic Geoscience Italy, Rome, Italy
e-mail: dimaggio@geologiaforense.com

© Springer International Publishing AG 2017
R.M. Di Maggio, P.M. Barone (eds.), *Geoscientists at Crime Scenes*,
Soil Forensics, DOI 10.1007/978-3-319-58048-7_2

2.1 Minerals

A mineral is a natural solid substance that has a characteristic internal crystalline structure due to the regular three-dimensional arrangement of its atoms. Its chemical composition may be well-defined or variable within very precise limits, and it manifests specific physical and chemical characteristics. The mineral may consist of a single chemical element, such as carbon (C) or sulphur (S), or more elements related in a simple chemical compound, such as quartz (SiO_2), or of compounds with complex chemical formulae including crystallization water molecules (OH o H_2O). Today there are over 4000 known mineral species, the classification of which is managed by the International Mineralogical Association (IMA), responsible for the approval and designation of new species found in nature.

Minerals are the fundamental constituents of geological materials, including those of forensic interest. They are present in the loose materials, such as sands and soils, as well as in rocks. Furthermore, most known gems are made of rare minerals with particular properties of beauty and hardness.

Minerals are essential materials in human activities for which they have always represented a primary resource. Some minerals are used as they are, such as talc, silver or asbestos; others are treated to obtain usable materials in the most varied fields of application. Consider their use in electronic and mechanical components, building materials and ceramics, paints and pigments, cosmetics, pharmaceuticals and so on. In addition, the crystal structures produced in nature throughout geological time represent a source of inspiration for undertaking synthesis, i.e. for the artificial reproduction of technologically advanced materials.

The study of minerals, their structure and composition, therefore, is of forensic interest not only for investigations involving the Geosciences, but also for all those criminal investigation sciences that have as their purpose the identification of every type of material, regardless of its organic, inorganic or synthetic nature.

On minerals chemical analysis are carried out aimed to the identification of their crystalline structure and the elements that compose them. Moreover, physical analysis are conducted to determine their main properties, such as hardness, color, luster, tendency to flake and/or fracture, birefringence, density, conductivity and any possible fluorescence, magnetism and radioactivity.

2.2 Rocks

A rock is an aggregate of minerals and/or amorphous substances (i.e. without a crystal structure, for example glass), naturally formed by diverse processes. Rocks are divided, according to their formation process, in three major groups: igneous rocks, which are divided into intrusive and effusive, sedimentary rocks, and metamorphic rocks.

Igneous rocks have a common genesis from the consolidation of high temperature fluid masses. The consolidation mode can occur deep within the earth's crust, giving rise to the intrusive rocks, or on the surface, giving rise to extrusive or volcanic rocks. Sedimentary rocks are formed on the earth's surface by exogenous agents. These rocks are generated, mainly, through the recycling of outcropping rock formation material, whose decay products are deposited through mechanical, chemical and biological processes. Metamorphic rocks are generated by high temperature and high pressure processes acting on pre-existing rock materials. Thus, this type of rock can be derived from the transformation of sedimentary, igneous or other metamorphic rocks.

In each rock, there are imprinted the processes that led to its formation and its characteristics; through its study, therefore, it is possible to trace the environmental conditions in which it was formed. There are thousands of types of rocks and various analytical methods can determine the comparability between samples depending on their genetic mode.

Rocks are geological materials that are often involved in crime narratives, both because they are present as fragments in the soil, therefore present as loose geological materials, and because they are frequently used as a blunt instruments or ballistic objects.

Usually, rocks are subjected to identification analysis for both the comparison of multiple samples, and for determining their origin.

Case Report

In the summer of 2002, on the eve of a Jewish religious celebration, in the Verano monumental cemetery in Rome, some Jewish graves were damaged: the tombstones were broken, some graves were opened and some coffins were uncovered. It was the first manifestation of racial intolerance at the monumental cemetery of Rome; it was feared that neo-Nazi groups had formed in the capital and that such acts were related to prevailing international political events related to the Israeli-Palestinian conflict.

Few days later, the investigators became suspicious of some gardeners who were working illegally in the cemetery. The Police inspected the cemetery and seized the gardening tools belonging to the suspected gardeners. Among the tools they found pickaxes and iron bars, on which there were traces of white material. The gardeners claimed they had been using those tools to do some maintenance work inside the cemetery, using lime and cement; however, there was a suspicion that gardeners had used the iron bars and pickaxes to damage the graves. For this purpose, samples of marble and travertine material were collected from the damaged gravestones and compared with the white material found on the tools.

Stereoscopic microscope analysis allowed for detailed examination of the morphology of the white traces that were of pasty, not compact, appearance, easily removable, and showed micrometric streaks whose orientation was compatible with a downward movement of the tool head (Fig. 2.1).

The white material traces on the tools and the stone samples were submitted for x-ray diffractometry analysis which allowed abundant calcite and secondary quartz and plagioclase constituents to be identified in all the samples. To rule out the presence of traces of building materials on tools, analysis were performed on different types of lime, mortar and cement, taken as references, which turned out to be mainly composed of calcite, portlandite and larnite, therefore crystalline phases

Fig. 2.1 Crowbar with traces of white material and streaks of this material viewed under the stereoscopic microscope

totally different from those present in the clue traces of white material typical of the gravestone samples.

For further confirmation, analyzes were carried out with a scanning electron microscope (SEM-EDX) to identify the chemical elements present in the white material traces in the stone samples and in building materials taken as references. The material on the tool and tombstones samples contained calcium in abundance and secondary quantities of aluminum and silicon, with the same composition percentages. The building materials contained calcium and silicon in abundance, but did not have aluminum.

These results provided evidence that the traces of white material on the tools had not been left by construction materials, such as lime, mortar or cement, but more likely by violent contact with the tombstones.

2.3   Sands

Sands areclastic compound deposits with at least 50% of the particles having sizes between 2 and 0.063 mm.

Virtually, sand can be found anywhere on the Earth's surface, but it is commonly associated with coastal areas, rivers, lakes, and where exogenous agents, such as water and wind, act frequently with sufficiently high energy.

Sand can be formed by the action of three main mechanisms: erosion of existing rocks, of any nature; chemical precipitation from supersaturated water into ions; accumulation of the skeletons and shells of organisms, such as seashells. In the first case, the composition of a sand, as well as its color, depends on the rocks that generated it and it is therefore closely linked to the catchment area of origin. For example, in the vicinity of a volcano, the sand will tend to be dark; near to a coral reef, the sand will appear white-pink; in non-tropical seas, it will mirror the composition of the rocks in the surrounding mountain areas (Fig. 2.2).

Fig. 2.2 Sample of white-yellow beach sand collected at a beach in South Sicily, where the surrounding lithotypes are mainly represented by deposits of marginal carbonate platform, and used as a basis of comparison with sandy clue traces on the footwear of a murder suspect

Local rocks are the source of most of the grains which make up sand, and as the sand is formed in different environments of the earth's surface, it can show very different characteristics from one place to another, with wide variability of composition and different micro textures and morphologies of the individual granules. For this reason, the study of the sand allows for its specific formation areas to be identified.

In general, the analytical techniques and the terminology adopted in the geological study of sands may be applied, except for a few exceptions, to any type of loose sediment, regardless of the size of the particles that constitute it.

The main processes responsible for the formation and evolution of the sand, as well as of all the loose geological materials, are erosion, transportation, sedimentation, and the early stages of diagenesis.

The erosive processes mainly include the alteration and the consequent disintegration of the rocks that, originated in specific chemical-physical conditions and brought to the surface by tectonic causes and/or erosion, undergo transformations that determine new physical-chemical balances aligned to changing environmental conditions. These changes can have physical, chemical and biological causes.

Physical processes include the fragmentation of the rocks and the subsequent reduction of the fragments and generally are linked to mechanical attack by exogenous agents, such as water, wind or ice, to the expansion of the rocks resulting in thermal variations and to structural and tectonic causes. Chemical processes take place by the chemical attack of water, whose solvent action may vary depending on its temperature, the pH (the concentration of H^+ ions) and the eH (redox potential). Biological processes are more limited than those described above, because they act in detail, and are related to the activity of some organisms that break down the rocks by means of chemical and mechanical processes in response to biological needs.

The transport of disintegrated erosion residue material can be effected by natural, chemical and biological means, and mainly includes: water related actions such as the transport of suspended matter, bed load and chemical effects, biological transport, as well as transport by the wind and glacial motion. Transport in suspension concerns the finer particles and is characterized by long stretches in which particles are suspended in a moving fluid (suspended load). Suspended load also includes transportation due to turbulence (saltation), where particles with intermediate dimensions are subject to leaps and creep proportional to their weight. Bed load affects the larger and heavier particles, which are put in motion by dragging or rolling. Chemical transport occurs by colloidal suspension or solution: the first relates to almost insoluble inorganic substances, the second involves all the other substances and depends on their solubility characteristics, the environmental condition and by the chemical characteristics of the solvent. Biological transport is operated by living organisms that remove chemical substances from the environment, and deposit them after their death in locations that can be quite distant from their place of origin. In wind transportation, the particles are predominantly transported by suspension as in the cases of volcanic ash, or by saltation, in the case of sands. In glacial transport, the transport medium is ice that mobilizes the debris, including large particle material, depositing it at considerable distances from its place of origin.

Sedimentation is the deposit and accumulation of organic and inorganic solid materials, on the Earth's subaerial surfaces or underwater areas. The sedimentation processes include: the precipitation of solutes from the water, gravitational processes and all the conditions in which the transport energy decreases until exhaustion.

Diagenesis includes the set of physical and chemical changes that occur within the sediment during and after its deposition. Such transformations are dependent on variations of pressure and temperature and to chemical exchanges with the surrounding environment. Diagenesis may lead to initial cementation of the sediment, then the formation of a compact rock.

Diagenesis in sands, and in the loose sediments, is generally very slight, i.e. it develops only at an initial stage where an interchange of chemical processes, which mostly alter the minerals, compared to the actual compaction processes, prevails.

2.4   Dusts

Dusts are generally constituted by a mixture of inorganic particles, single or in aggregates, with dimensions of silt (lower than 63 μm) and clay (less than 4 μm) and organic particles.

Natural dusts can be produced by any geological surface devoid of coverage. The deserts and semi-arid regions, are the main concentration locations for the world's dust, but also important are dirt roads, uncultivated land and sand banks can act as sites of local dust production. Other sources of natural inorganic and organic dust are represented by volcanic activity, fire, fields and forests during periods of flowering, pollen-producing powders which, despite being organic, blend with inorganic and/or anthropic dusts.

Anthropogenic dusts can come from industrial environments, such as foundries and construction sites. In the air of urban environments, there are particles mainly derived from vehicle emissions, from domestic heating plants and, albeit to a lesser degree, also wear debris from brakes and tires.

Dust particles are usually transported suspended in the atmosphere. Particles with size greater than 20 µm tend to precipitate by gravity rather quickly and will not be transported for long distances, unless they have a particularly low density. The finer particles, with sizes less than 10 µm, can remain suspended in the atmosphere for long periods and may be transported over great distances, even of the order of hundreds of kilometers.

The traces of dust, both natural and anthropogenic, due to the high mobility of its particles, can be found on a wide variety of surfaces of forensic interest, thus, in some cases, they can be considered as valid characterizing element.

Dust can be of forensic interest also in cases of air pollution: dust particles with diameter less than 10 µm, so-called PM10 (Particulate Matter), in fact, can reach the different parts of the respiratory apparatus, also causing severe health problems.

2.5 Soils

Soil is a natural body with of a solid component, distinguished in organic and inorganic parts, with a liquid component, mainly composed of water and dissolved salts, and a gaseous component, formed by air which is generally richer in carbon dioxide and poorer in oxygen.

Soil is differentiated according to variable layer depths that differ in underlying material morphology, physical constitution, properties and chemical composition and biological characteristics. The soil is concentrated in the upper section of the regolith (i.e. the set of loose materials that form a mantle that covers the underlying solid rocks) it covers the earth as a thin layer and has the function of support and nutrition for vegetation. The soil is the result of a complex set of chemical-physical, biological and climate factors that determine the characteristics of the evolutionary process.

Soils have highly variable chemical and mineralogical, and grain size-textural compositions, depending on the characteristics of the substrate rocks, their degree of alteration and the weather conditions.

The soil formation involves long chemical, physical and biological alteration processes occurring within an accumulation of disintegrated and non-consolidated material, resulting from alteration of some rock types, simultaneously and after its deposition. These processes involve the alteration of inorganic and organic compounds, their deposition and the subsequent formation of new minerals and new organic molecules. The pedogenetic substrate, from which the soil evolves, is formed mainly by the processes of erosion, transport and sedimentation. The composition and structure of a soil depends on many factors, including: (i) the nature of the parent rock, i.e. the physical, mineralogical, and geochemical origin of the rock, which is responsible for the alteration tendency of a soil; (ii) the climate, which is considered the main responsible agent for the formation and the definition of the properties of the soil because it includes changes in temperature, precipitation patterns, and the shaping action of water; (iii) water, which affects the majority of the physical, chemical, and biochemical processes during the formation of the soil; (iv) the activity of biotic entities, namely the vegetation and the micro and macro fauna; (v) the topography and morphology, defined by the altitude and the slope of the area; (vi) time, as the different products and the definition of the alteration of soil characteristics occur in longer or shorter times; (vii) human activity that can alter the direction and speed of soil formation through targeted land use actions.

The set of processes that lead to the formation of soils also involves their division into horizons, namely vertical differentiations which are all the more evident the more the soil is evolved. From top to bottom, a mature soil consists of a cover layer, called horizon O, and four horizons, named A, B, C and R. The horizon O is the most superficial layer, of limited thickness, composed of organic substances that are decomposing or not yet fully decomposed. In this layer the mineral component is absent or very limited. The horizon A is the richest layer of organic substance, in which the decomposition activity is very evident and where water, which moves from top to bottom, sustains the removal of the soluble salts and colloids. The upper level horizon A_1, is characterized by the presence of humus-based organic material, which gives it a dark color; the lower level, called A_2, represents the area of maximum leaching of colloids, salts and organic matter, which results in a lighter color compared to the level above. The B horizon is poorer in humus than the horizon A, but is also the layer that represents the area of maximum accumulation of substances leached from horizon A. In this horizon are concentrated clays, iron ores, carbonates and humus, making it more defined and with more pronounced color compared to the others. The C horizon is not part of the soil in the strict sense and consists of the little altered fragments of the parent bedrock towards which it tends to revert. Finally, horizon R represents the unaltered underlying bedrock. The thicknesses of the horizons and of the covering layers can be very variable, from a few millimeters to meters (Fig. 2.3).

2.5.1   The Classification of Soils and Soil Maps

Soil classification is concerned with the systematic characterization of the soils which can be based both on their properties and on the parameters which determine their use. The classifications based on the distinctive characteristics of the soils can refer to the climate in which they are formed or the lithology of the substrate, pH or their structure.

In recent years, further classification criteria have been developed that are based on land use criteria. For example, agronomists tend to classify soils according to their chemical-physical properties as appropriate to particular types of cultivation; geotechnical engineers, classify soils according to their engineering properties in order to assess their potential for sustaining building foundations or as a building

Fig. 2.3 Measuring the
level of the horizon in a
soil developed in a coastal
dune environment

materials. Soil classification systems are based on specific technical requirements,
they also set the evaluation criteria and recommendations that guide the choices for
optimal land use and management.

These classifications have little relevance in the forensic context; in fact, they do
not involve detailed micro-environmental soil differentiation, such as those minimal
differences in soil properties that may prove extremely useful in the study and comparison of samples for judicial purposes. Forensic soils cannot be easily classified
also because they are generally represented by samples that have been handled and
modified with respect to the soil of origin, and sometimes there is very little volume
of these soil available.

The soil maps are graphic representations of the geographical distribution of
soils and they are developed according to specific classification schemes. Detailed
soil maps are drafted according to the scale of the soil survey and cartographic
rendering techniques. The greater the map detail, i.e. the larger the scale, the more
detailed are descriptions of the properties of the relevant illustrated soils and the
more accurate are boundaries between different soils. The scale of detailed soil
maps can be used in forensic applications where there is the opportunity to pinpoint
areas with soil characters like those of a trace sample of soil for which it is necessary to determine the probable origin.

2.6 Gems

Gems are special species of minerals and, much more rarely, rocks, that, thanks to their beauty, hardness and rarity, have an extremely high monetary value. In general, a precious stone is defined as mineral from which a gem can be extracted, the cut and finished gem itself is then considered as a mineral often used in jewelry. There are also gems of organic origin, such as amber, coral, pearls and ivory. Since Neolithic times, mankind has vested particular interest in precious stones, using them not only as precious ornaments, but also as symbols evoking magical and spiritual virtues. Gems and precious stones have never ceased to fascinate; all peoples throughout the ages have left evidence of their interest in the beauty and use of these natural materials.

Unfortunately, gems and precious stones have not attracted man only for their beauty; their high economic value, small size and their non-traceability have made them are also extremely attractive to organized crime. Because of the traceability of bank movements at national and international levels, criminal organizations are using precious means of exchange for the traffic in arms, drugs, and for all those crimes which require a monetary transaction. Gems, as well as jewels, precious metals and works of art, are used for money laundering and by criminal associations as an alternative form of investment of illegal proceeds.

Besides the crimes listed above, there are also those relating to the fraudulent falsification and imitation of gems. Because of the rarity of precious stones, the practice has evolved of resorting to various kinds of falsification. Already in ancient times, it was customary to falsify gems; the earliest evidence of changing the appearance of gemological material dates back to the Minoan era. Over the centuries, with the new knowledge of materials and development of chemical and physical techniques, artificial alternatives based on lower value materials have evolved as replacements of gems.

The following are the main features of the gems to understand the types of forgery and imitation.

2.6.1 The Characteristics of the Gems

Color is one of the characteristics that allows for the identification of a gemstone and an estimation of its value, although gems that belong to a same mineralogical species may have completely different colors. Color is also one of the most easily forged features and color, alone, should not be a criterion for gem identification.

By their nature, gems have distinctive signs, which may be superficial, called surface signs, or internal, known as inclusions. Inclusions are formed during the mineral crystallization process and they do not necessarily diminish the beauty of the gems. The purity of a gemstone relates to its degree of internal cleanliness, the nature, number, arrangement pattern and the size of the solid and liquid inclusions, and any structural distortions caused by the crystal growth process. The degree of purity is based on what is visible under ten-fold magnification. Usually, the greater the purity of a gem, the greater is its value, especially in diamonds; there are, however, cases where inclusions that do not interfere with the hardness of the gem, can also make them more beautiful and more valuable.

The carat is the unit of measurement of precious stones and is equivalent to a fifth of a gram, thus 200 mg. The carat is an ancient unit of measurement which owes its name to the traditional use of carob seeds to weigh gems, especially in the Middle East. The seed of the carob tree is called *qirat* in Arabic, a term that derives from the Greek *keration*; hence the name carat. In general, other characteristics being equal, the greater the weight of the gem, the greater its rarity, and, therefore, the higher its value per carat. For example, a three-carat diamond will be worth much more than the sum of three diamonds of one carat each of the same quality.

There are many types of gemstone cuts according to aesthetic preferences and the type, size, and the original shape of the stone to be cut. Generally, the choice of the cut should be made so as not to alter the degree of purity and the proportion of the gemstone, as well as prevent the loss of significant quantities of material. The cut of a gemstone is determined by three elements: brightness, proportions, and finish. Depending on the combination of these elements, the cut is classified as excellent, good, standard or mediocre. The brilliance is directly related to the reflection of light inside the gem; the proportions represent the relationships between the angles and the various measures of the gem; the finish is the set of the quality of polishing and cutting accuracy. The finish depends on the external characteristics of the stone and the type of cut chosen.

2.6.2 The Falsification of Gems

Falsification of gems takes place mainly by three main methods: the exaltation, the synthesis, and the replacement.

The exaltation of the gems is a practice that allows for the embellishment of precious stones, exalting the characters already present or hiding defects commercially under-appreciated, in order to artificially transform a poor gem in a more prized specimen.

Exaltation is a practice already in use for several centuries, it is mostly performed to improve or change the color of the gems. Today there are many treatments for enhancing the color or transforming a colorless stone in a colorful one, using techniques of heat treatment, irradiation or by actually applying colors. Other types of exaltation involve procedures designed to hide clearly visible flaws, such as fractures or cracks. The main actions consist of filling cavities or crevices with fluid hardening materials, in total or partial coating of the gem with foreign substances or the permeation of fractures with oily liquids.

The synthesis of gems is their artificially reproduction them with specimens having the same chemical, physical and optical properties of the corresponding naturally occurring items. There are different synthesis processes, however, the most common one consists of bringing to melting a composition of material equal to that of the mineral that you want to imitate to which are added metal oxides which give it the desired coloring; subsequently, the mass of molten material crystallizes as a function of slow cooling.

The synthetic stones are quite recognizable under the microscope as they exhibit typical curvilinear striae of growth and sometimes gas bubbles. In recent times, however, the methods of synthesis have reached such advanced techniques to obtain products almost perfectly identical to their natural corresponding to the point of not being able to easily distinguish the differences. In these cases, to definitive identify the product of synthesis, very advanced and expensive instrumentation is often used.

Another counterfeiting method consists of replacing the most precious gems with relatively common natural stones of lesser value or with artificial materials, such as glass paste, ceramics or plastics resins.

Finally, a technique that allows saving or replacing precious material with low-quality material is that of the doublet. It consists of creating a gem consisting of two parts; the upper one is normally a precious stone which is affixed to a lower part of lesser value. Sometimes, doublets of very precious stones are made exclusively with less precious materials, as for example the ruby that is often imitated with a red garnet and glass doublet.

The analytical methods used for the study of the value of gemstones consist of the measurement of refractive index, specific gravity, hardness, the caliber and the study of all significant characteristics, such as color, clarity and cut. In cases of imitation with synthetic products of last generation and to provide reliable indications as to the genuine or false nature of a gem, advanced chemical and physical analysis on the gems are carried out to study the characteristics in detail, not determinable with superficial analysis or low power magnification.

2.6.3   The Main Precious Gems

Diamond is the most sophisticated and the most precious stone used in jewellery in all parts of the world. Given its application also in the technological sciences, we can say that nine-tenths of world trade in precious stones are represented by the diamond and thus, such ratios remain valid also for its use in the context of criminal activity. The diamond evaluation criteria are based on the purity, the color, the cut and the weight. The degree of purity and the color of the diamonds are determined according to scales established by the Gemological Institute of America (GIA) and used worldwide.

Forgery and imitations of this gemstone are numerous. In several countries, for several years, synthetic diamonds have been obtained starting from graphite subjected to high pressures and high temperatures; however, these procedures have higher costs than the natural rough diamond. In recent years, a new low-cost synthesis product called *moissanite* has undermined the diamond market. Synthetic moissanite has hardness and characteristics very similar to those of diamond and has a value of approximately 10–15% of a comparable quality diamond. However, the brilliance and the specific gravity are lower compared to the diamond; also, moissanite it is birefringent, while diamond is monorefringent. Another synthetic product that is commonly used to replace the diamond is the cubic zirconia. Often the imitation of a diamond is carried out by replacement with lower value gems, whose physical-optical properties are similar to those of diamond, or even with fine glass.

Contrary to the diamond, colored stones cannot be subject to profiling about the color grade and purity; the value of these gems is based, however, on a number of features such as color intensity, hue, transparency, brilliance and cut.

The most valuable gems, after diamond, are corundum and beryl. The corundum has a very extensive range of colors, but the red specimens, rubies, and blue sapphires, are the most valuable pieces from the point of view of gemology and also the most counterfeited. Ruby and sapphire have been among the first gems to be synthesized, already in mid-nineteenth century; however, only in the early years of the twentieth century were synthetic copies of acceptable quality obtained through the fusion of powder of aluminum oxide to which a metallic dye was added (Fig. 2.4). Corundum can also present themselves as reconstituted specimens, obtained by melting natural fragments of these stones; the result is a product in the glassy state, of poor hardness, that has nothing to do with the real synthesis (Fig. 2.5). In attempts to imitate with less precious stones, the ruby is generally substituted with the red spinel and tourmaline, while sapphire is substituted with the blue spinel and kyanite.

Beryl is a very common mineral that has always played an important role in gemology for its magnificent green specimens in emeralds, blue and aquamarine colors. Because of its high value, emerald often undergoes attempts at falsification by hydrothermal or synthetic copies and is imitated by replacing it with less valuable stones such as green spinel, fluorite or quartz-prase. From the points of view of demand and market deployment, the aquamarine is the fifth most important stone after diamond, ruby, sapphire and emerald and has always been the subject of forgery and imitation. The least valuable stone with which it is replaced is generally the blue topaz, especially since it has been possible to produce them in synthetic form in large quantities, but are also used are blue quartz, the synthetic sapphire, blue zircon and glass pastes.

Numerous are the precious stones of lower value to those mentioned previously, but which also are falsified with synthetic materials, plastics, resins or imitated with stones of the same species, but poorer, or with minerals of lower value and glasses. Among these, for example, jade is falsified by means of impregnation of pale color specimens so as to impart to them an appearance similar to the more valuable green jades. Jade imitations are represented by similar less valuable minerals, such as amazonite or quartz. Other stones are frequently falsified include opal, garnet, tourmaline and topaz.

Organic gems such as pearls, coral, ivory and amber, which are not among the colored stones, are also frequently subject to numerous attempts of falsification. For example, in recent years, amber, which is an interesting gem that can have an age between 10 and 45 million years, in addition to being falsified with polymerized synthetic resins, is being substituted with natural non-fossilized resins which are heated and subsequently pressed to give them a more aged appearance.

Fig. 2.4 Synthetic
sapphire, dating back to
the early twentieth century,
mounted on gold bracelet

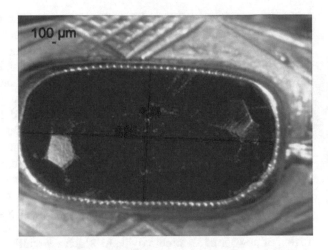

Fig. 2.5 Reconstituted
rubies mounted in gold
into a jewel of the 1940s

Case Report

One of the most popular precious gems scams of recent years, perpetrated in various Italian regions to the detriment of many elderly people, was the so-called "Sailor scam", albeit with different variations on the theme. It was staged by a gang of three fraudsters: an apparent foreign sailor, a fake translator and a goldsmith, ready to assess the goods. The latest scam of this type has been perpetrated at the expense of a 64 year-old man in the province of Trapani.

The victim was approached by an apparent French sailor in his 30s (first swindler) who asked him how to reach the local information office. The elderly gentleman, unable to help the sailor, took advantage of the presence of a passer-by (second swindler) to ask for directions and he said he knew where the office was and offered to help the French sailor in distress. The sailor asserted that he needed to reach that office in order to sell some diamonds to earn money to return home. The passer-by, intrigued by this fact, interjected into the conversation, and, claiming to know a local

Fig. 2.6 Specimens of cubic zirconia seized from the counterfeiters who used them for the "Sailor scam"

jeweler (third scammer) capable of estimating the value of the jewels, he walked away to call the jeweler and then returned after a few minutes in the company of this intermediary. The fake jeweler, with a professional air, invited those present to discretely enter his car to close the deal, according to him, the value was at least 14,000 Euros. At this point, the sailor said he would also settle for a lower amount. The passer-by then proposed to the sailor a bid of 7000 euro, but he did not have the cash on him at the time. The victim, taking advantage of the lack of liquidity and through fear of losing a great bargain, reached an ATM, withdrew 1500 Euros closed the deal for himself. On returning home the elderly gentleman became suspicious and, after an assessment of the goods, he realised that instead of purchasing of precious diamonds, had spent a huge sum for zircon stones worth about 15 euro each (Fig. 2.6).

2.7   Precious Metals

The metals are elements distinguished by high thermal and electrical conductivity as well as for certain physical properties such as hardness and fusibility. Depending of their chemical stability, they are distinguished in precious metals and base metals.

Precious metals are particularly resistant from the chemical point of view and possess particularly important physical characters, such as for example ductility, i.e. the ability to withstand plastic deformation without rupture. In the commercial field the well-known precious metals are gold, silver, platinum and palladium.

Objects produced with precious metals, or alloys of these metals which have a legal title, have become a new form of investment and money-laundering. This aspect has increased dramatically the illegal trade in and smuggling of precious metals. The illegal trade of these materials mainly involves countries such as China, Russia, and many tax havens.[1] Together with the gold smuggling, there is a growth in money laundering and receiving of stolen precious metals through numerous "We Buy Gold" stores opened by organized crime syndicates, which, in addition to laundering dirty money earned from drug dealing and pimping, perpetrate frauds to

---

[1] Into Italy, over about 4 years (2008–2012), one and a half tons of gold worth over 26 million euro were smuggled.

the detriment of customers by subtracting several grams to the actual weight of the precious material.

The crimes listed above are joined by those relating to the falsification and the imitation of precious metals and their alloys in jewelry making, bars, coins, and other objects involving criminal intent.

2.7.1 The Character of Precious Metals and Their Alloys

Pure precious metals are too soft and malleable for the production of jewellery and objects, for this reason they are alloyed with other metals that improve resistance and workability, also with a change color depending on the aesthetic choices; for example, the different shades of gold depend from the different percentages of the metals that are added to pure gold. Generally, gold is alloyed with copper, platinum and/or nickel, silver with copper or zinc, platinum with iridium, palladium or copper.

Since a superficial examination cannot verify the precious metal content in an alloy, Italian legislation prescribes that objects made with precious metals and their alloys should be compulsorily inscribed with at least two signs that constitute the so-called hallmark: the indication of the purity and the identification mark.

The purity is expressed in thousandths and highlights the relationship between the mass of the metal alloy base and the pure or fine precious metal. For example, an 750 ‰ alloy will contain 750 fine metal parts and 250 parts of other metals. Purity expressed in carats but is no longer used, although for gold it is still widespread.[2]

The hallmark is represented by a polygon containing a five-pointed star, a unique number assigned to the manufacturing company by the Chamber of Commerce and the initials of the province where the company resides. The hallmark allows the manufacturer to be traced and, therefore, retained as legally responsible for the purity compliance declared for the item (Fig. 2.7).

2.7.2 The Falsification of Precious Metals

Generally, counterfeiting of precious metals occurs by substitution with less precious metals and subsequent plating with the metal to be imitated, as well as employing different alloy proportions compared to the specification. Also, the trade mark and the purity mark are subjects of counterfeiting.

Typically, the falsification of native metals samples is quite frequent because they belong to the species of considerable commercial value, in fact, the value of the piece exceeds by several times the value of the precious metal content. The falsification of these specimens is quite easy: the fakes are made from alloys of the metal subject to imitation; for example native gold with a gold alloy.

---

[2] 18-carat gold is an alloy that contains 18 parts of pure gold out of 24, where 24 carats refers to the pure metal.

Pure gold ingots, weighing 400 oz, are the subject of numerous fakes, both by forgers and governments, as they constitute the standard of interbank trading. Years ago, the process used to falsify gold bars was to add copper to gold, and then to make an alloy, or to laminate with gold an equivalent volume of steel or lead. Steel and lead, however, have a specific gravity less than that of gold, thus the pieces forged with this method were weighed 60% less than the genuine version, making them easily identifiable as fake.

In the last decade, several counterfeit gold bars were discovered, forged by the coating of tungsten blocks with a thin plate of pure gold. The choice of tungsten proved to be very advantageous because it has a monetary value far less than gold, but a similar specific weight, so it is virtually impossible to distinguish a gold ingot from a tungsten forgery only by the weight/volume ratio. This forgery technique has provided the idea of the counterfeiters who produced tungsten ingots using different methods (Fig. 2.8).

Fig. 2.7 Engraving of the purity and the hallmark on a silver platter

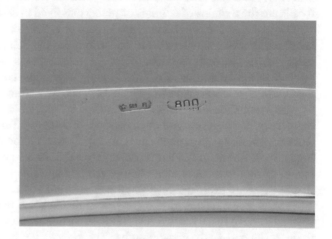

Fig. 2.8 Gold bullion with tungsten core. The forger has drilled a genuine bullion and filled the holes with tungsten (Source: ABC Bullion)

The risk of falsification also involves silver bullion, in fact, that metal can be replaced by molybdenum, both having very similar specific gravities. Although fake silver bullion still has not been found, recently a marked increase in the production of molybdenum has been noticed.

2.8   Fossils

Fossils are the remains or traces of organisms both plants and animals that have been preserved in sediments that later consolidated into rocks. When an organism dies, generally, its hard parts are preserved, while the soft parts decompose very quickly. For the process of fossilization to occur, is necessary that the body be buried rapidly and this happens more frequently in underwater environments where sedimentation processes are faster than those of decomposition; for this reason, the most common fossils are the remains of animals or plants that lived in seas and lakes.

The process of fossilization is a very rare phenomenon, which needs a considerably long time and which depends on various physical-chemical processes that take place within the sediment. The most common process of fossilization is that of mineralization, by means of which the body's chemical composition is modified by the action of solutions which circulate between the sediments.

Many specimens of fossils have achieved considerable commercial value, due to their rarity, the conservation status and their antiquity. Parallel to the huge market demand for these specimens, their manipulation and/or falsification has become increasingly common and has improved over time to the point of reaching such high levels, thanks to modern technologies employing tools and materials of all kinds, that in a general examination, even the experts are misled.

Parallel to the falsification of fossils, in recent years there has developed a contraband trade in fossil records involving, in particular, dinosaur skeletons or parts of them, tusks and teeth of mammoths. The clandestine nature of the smuggling of precious fossils makes it impossible to calculate the dollar amount of world trade, but reliable industry expert estimates suggest that it is about tens of millions of dollars a year.

In Italy, any activity related to paleontology is bound by the laws of the Ministry for Heritage and Cultural Activities, governing paleontological research and activities, that prohibit the collection to anyone who does not provide serious and adequate guarantees.

2.8.1   The Falsification of Fossils

False fossils include both pieces reproduced more or less faithfully with various materials and that those which have undergone a partial reconstruction or a touch-up, the addition of material details unrelated to the original fossil, or even the recomposition in positions different from the original.

Fig. 2.9 Fake fossil specimen with the position of the fish discordant with respect to the stratification (Source: www.terranea.it.)

The counterfeiting of fossils aims to improve their appearance to entice a buyer to purchase; for this reason, all fossils with commercial value are often subject to manipulation, although the more falsified fossils are trilobites, ammonoids and fish, because they are the most popular (Fig. 2.9).

The main forgery methods include the reconstruction of missing parts, the positioning of the fossil on a piece of matrix, the assembly of pieces belonging to various individuals, artificial coloring, and total reproduction. The latter is usually carried out by means of a mould that can be produced using various materials; generally, appropriately colored resin is used. Resin is also used to make partial manipulations of the fossil. For example, during recovery operations, some types of fossil tend to detach from the matrix, as occurs, for example, with shark teeth; in these cases, the tendency is to re-glue with the resin the pieces that broke away, thus creating a falsification of the original situation.

The methods used for the study and appreciation of the fossils are the essentially paleontological observations which include the identification and classification of the test specimen, the taphonomy, i.e. the study of processes ranging from the death of the organism to its fossilization, and, in the case of most specimens in the finding, the association of species. In cases of falsification, further chemical analysis can be undertaken to determine the nature of the materials that compose the finding, as well as dating using the radioactive isotopes of carbon 14.

2.9   Microfossils

Microfossils include all the microscopic organisms (smaller than 1.2 mm), and/or parts thereof, which have been preserved in the sediments, and have fossilized over time. The science that studies microfossils is micropaleontology in whose field of study are included protozoa (including foraminifera, planktonic and benthic, radiolarians, ostracods, namely microscopic crustaceans), algae, such as diatoms, pollen, the spicules of sponges, coprolites, and fish scales. All these organisms or parts of them, due to their small size and their cosmopolitanism, are found in practically all sedimentary deposits and can also be present in large numbers in a rock and then also in the ground from which it is formed.

Microfossils play a considerable part in the study of soils and rocks in the forensic context as their presence in different geological materials, can provide useful information for the characterization of the sample to the point of also establishing its probable origin. For example, the populations of microfossils present in soils vary depending on the deposition environment of the sediments from which the soil is subsequently formed.

The presence of microfossils, their variety and/or their particular associations, can provide additional information on the characteristics of an area or on the location of a site. The associations of microfossils, in addition to having substantial variation by area, undergo remarkable changes both in qualitative and quantitative terms, also between stratigraphic levels of the same formation. In other words, passing between one level and the other of a same stratigraphic unit, it is observable both the fossil species mutation and the mutation of the percentages of individuals of the same species.

Microfossils present in soil or rock are normally isolated and concentrated during laboratory operations, and later studied under an optical or electron microscope for identification and classification purposes.

Chapter 3
Pedology Applied to Forensics

Rosa Maria Di Maggio

Abstract Pedology forensic research studies and analyzes traces of geological debris and loose materials connected to a crime, in order to provide assistance to the judicial police activities and to obtain evidence useful in carrying out an investigation. In the context of forensic analysis, such materials are commonly referred to by the term 'soil' and include sand, mud, soil in the strict sense, etc. This chapter treats the characteristic of forensic soil and the issues related to the traces of soil. The site survey activities in forensic soil science applications and the collection of soil trace evidence and their preservation are also dealt with the following pages.

Keywords Forensic pedology • Soil • Trace evidence • Site survey

Pedology is a branch of the earth sciences that treats soils in their natural environment; studies its genesis, morphological, chemical and physical characteristics, its classification and large and small scale geographical distribution.

Pedology forensic research studies and analyzes traces of geological debris and loose materials connected to a crime, in order to provide assistance to the judicial police activities and to obtain conclusive evidence useful in carrying out an investigation. In the context of forensic analysis, such materials are commonly referred to by the term 'soil' and include sand, mud, soil in the strict sense, etc..

Soil provides important information to link a person to a place and to clarify the narrative of a crime, because it can be transferred from one place to another by simple deposition on mobile surfaces, such as shoes, tires, floor mats of vehicles or tools (Fig. 3.1). The basis of this assumption lies in the important Locard exchange principle which states that there is always material transfer whenever two objects come in contact with each other.

Many crimes are committed in outdoor areas where the environment, which is not a closed system, can exchange items with those who are present in it; so soil can easily come into contact with the author of a crime or his victim. On the clothing, footwear or the car of a culprit, can be deposited grains of sand from the crime scene; on work tools can be found soil particles that can be traced back to the place where

R.M. Di Maggio (✉)
Geoscienze Forensi Italia® – Forensic Geoscience Italy, Rome, Italy
e-mail: dimaggio@geologiaforense.com

© Springer International Publishing AG 2017
R.M. Di Maggio, P.M. Barone (eds.), *Geoscientists at Crime Scenes*,
Soil Forensics, DOI 10.1007/978-3-319-58048-7_3

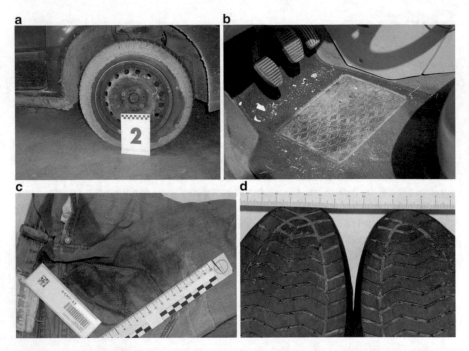

Fig. 3.1 Traces of soil present on different types of mobile surfaces related to different crime narratives: (**a**) mud deposits on the fender and tyre of a car involved in the running over of a pedestrian; (**b**) soil and other materials deposited on a car mat of a vehicle; (**c**) soil traces on the trousers of a victim of homicide; (**d**) soil traces on the shoes belonging to a man suspected of sexual assault

weapons and/or explosives were buried; at the scene of a crime can be found traces of mud, extraneous to the environment, that may indicate the origin or location of a likely offender. By way of example, consider the case where a murder victim was transported somewhere other than the place where he was killed; the analysis of soil samples may provide useful information about the relationship between the victim, how it was transported, the place where the offence was committed and a suspect.

Soil is made up of a complex system composed, at microscopic level, of several components that have been formed and developed during its long process of formation. This process is affected by many variables such as the nature of the parent rock, climate, topography, erosion, depositional environments, microorganisms and vegetation, tectonics, the hydrographic structure, and time. To the above processes, others are added due to human intervention and with actions, operations, and activities that change the structure, composition and chemistry of the soil to make it as suitable as possible for a particular purpose; consider, for example, agricultural soils that are enriched in nutrients to optimize their productivity. Human activity affects soil components also with the contribution of various materials, such as fragments of plastic, glass, brick, paper, that are simply deposited by gravity with concentrations that are in function of the time-space anthropic characteristics.

Because of this complexity, there may considerable differences between one sample and another, to the extent that considerable variations may be recorded in the components and their proportions, in areas of limited extension, in which the soil is considered to be homogenous. Careful analysis of soil traces allows for a characterization of the samples to the point that it is possible to distinguish whether they were collected a few meters away. For such evidence, the use of soil analysis is an excellent support to judicial investigations.

Generally, the forensic soil scientist makes a comparison between soil traces from the scene of a crime and those found on various objects belonging to a suspect or a victim. It may happen that, in the absence of comparison samples, examination is performed on the composition of the soil traces, assessing their likely geographical origin and, if appropriate, comparing them with samples of soil taken from a specific area, with known chemical, physical and geological characteristics. The conceptual, analytical and procedural approach of the forensic soil scientist is different from that of the soil scientist. In fact, the context and the purpose for which the analysis is carried out on the land, the type of samples and their limited quantities lead to the adoption of analytical and processing procedures on the data different from those used in the classical soil science.

The characteristics of the forensic soil scientist are indispensable given the fact that the variety of existing soil is endless and that, as with all the natural elements, it is impossible to find identical samples. The forensic soil scientist, therefore, analyzes and evaluates the level of comparability between soil samples (soil evidence); furthermore, by exploiting appropriate geological knowledge, he is able to determine the possible origin from the same geological or pedological micro-environment. In this context, the term micro-environment refers to an area of limited extension in which the soil does not manifest significant changes in its characteristics, its constituent elements, and their reciprocal profusion. For example, a micro-environment may be the garden of a house or a portion of a wooded area. The delimitation and thus the extension, of a micro-environment is determined by studying the variability of the soil characteristics within a specific area.

The complexity of the work of a forensic soil scientist is testified by a stage outside the laboratory, when he interprets and processes, at the crime scene, the geological, soil and environmental conditions that guide the location of suitable sites for the collection of soil samples. It is followed by a phase in the laboratory where samples are analyzed and their level of comparability are determined, through the study of the data obtained.

Case Report In December of 2011 the partially charred body of a Nigerian girl enslaved in the prostitution racket between Nigeria and Italy, was found in the Misilmeri countryside near Palermo. In the very early stages of the investigation, a man was suspected who, according to witnesses, had been the woman's last client.

During the autopsy, soil traces were found on the soles of the victim's partially charred shoes (Fig. 3.2a). During the inspection at the corpse discovery site, aimed at collecting evidence collection of soil samples and other clue traces useful to the investigation, the presence of some dried papyrus roots with soil aggregates was noticed (Fig. 3.2b). The presence of papyrus parts near the body of the victim was peculiar because in that area the crops and natural vegetation were different. The inspection and evidence collection of clue traces was also undertaken at the home of

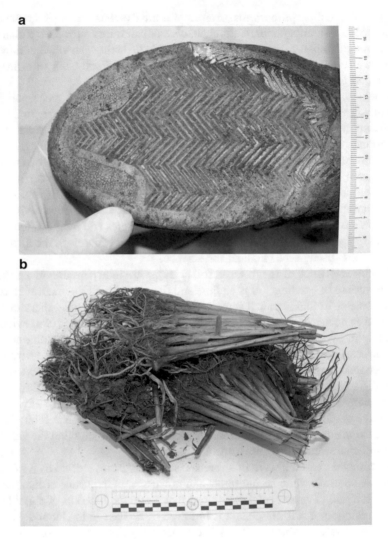

Fig. 3.2 (**a**) The victim's partially charred shoe with soil; (**b**) the roots of papyrus with soil aggregates

the suspect. In the garden of his house a pile of dried papyrus was found; samples of the papyrus roots presenting soil aggregates were taken and soil sampling was carried out both inside and outside the property.

With investigative information gathered to that point, a crime hypothesis was established: the man after bringing the woman in her home had killed her there; then he loaded the body on his off-road vehicle, covering it with dried papyrus and, after having dumped it in the countryside, set fire to it.

The purpose of the soil analysis was to compare the soil samples taken from the home of the suspect with the ones on the victim's shoes, for the presence of the woman in that place, and to compare the soil on the papyrus roots found near the corpse with that present on the roots in the garden of the suspect's home, to determine their same origin. If tests would have shown high comparability between samples, one could substantiate the crime narrative assumed by investigators.

Studies on the color characteristics of the soil samples and the size distribution of their particles showed very similar features to those taken from the garden of the suspect and those found in the victim's shoes. The strong similarities were also verified between samples of the papyrus roots found in the garden and in the vicinity of the victim, while the soils found at the corpse discovery site and outside the property of the suspect showed significant differences.

X-ray diffraction analysis revealed that the soils related to the papyrus, the footwear of the victim and the suspect's garden contained the same mineral species, such as quartz, calcite and kaolinite. The observation of thin-section soil samples under a polarizing microscope confirmed the high similarity between these samples and allowed for the extraction of more mineralogical information in support of the common origin of all the samples.

The stereoscopic microscope examinations (which simply enables a magnification of the sample to carry out detailed observations) made it possible to recognize the presence of the same types of particles in all the findings. Through detailed microscopic study it was also possible to detect the presence of very small dark red fragments in the soils taken from both papyrus roots, which seemed to be man-made. If these peculiar and rare fragments had shown the same chemical composition, together with the high comparability between the inorganic components of the soil samples, it would have supported a strong link between the suspect and the place where the body was found. Detailed chemical analysis were carried out with FT-IR spectroscopy, they helped to establish that both the dark red fragments taken from the papyrus roots consisted of clay paste for building purposes and alkyd paint. These results suggested that the fragments could be from construction materials such as roofing tiles.

The papyrus roots found near the corpse in the garden of the property and the suspect were analyzed by an expert botanist who, with the comparison of fresh papyrus samples collected in the same garden, identified both as *Cyperus Alternifolius*.

The data obtained from the analysis of the soil, of the particular plant species and anthropogenic materials, showed a high level of comparability between the samples and allowed to prove a strong link between the suspect and the place where the woman's body had been found. These results, along with investigative information obtained by investigators, corroborated the crime narratives assumed in the early stages of the investigation.

3.1 Characteristics of Forensic Soils

Forensic soils include all types of surface covering the earth's crust, formed by loose elements, whose particles can move on various surfaces, providing clue traces as possible sources of evidence in forensic applications. Forensic soils include soil, sludge, sand, scale, dust, from different places and environments, or found on clothing, vehicles, tools of various types and on the human body. So, the expression forensic soil refers to the loose material traces, of geological origin, present in the context of a crime rather than a definition based on the intrinsic characteristics of the material itself.

The forensic soil scientist usually has to deal with handled soil samples (soil traces) that are removed from the place of origin. In fact, the soil traces subjected to forensic analysis can be found in places where it is suspected that they had been transported by vehicles or shoes, or on objects that had removed them during the execution of the crime.

Generally, soils affecting forensic investigations are those that come from the upper layer of an loose deposit zone and, in general, may involve depths of no more than 3.5 cm. In fact, the most superficial particles are the ones that transfer to shoe soles, tires and clothing. It should be noted that there are cases in which, for the specific crime narrative, it may be necessary to collect soil samples also at greater depths. These are cases where, for example, shallow holes have been excavated to hide bodies, weapons etc. In cases where holes have been dug to conceal bodies or objects, a stratigraphic study of the concerned soil levels must be performed at the excavation site and samples must be collected at different depths and soil horizons.

The soil characteristics can change significantly over short distances and variations of depth; it is necessary, therefore, during the inspection, to have an actual or presumed crime narrative, to collect samples suitable for the applicable context.

Forensic soils are composed of three types of components, present with strongly variable reciprocal ratio: the inorganic component, the organic component, and the anthropogenic component (the set of various materials whose presence in the soil is attributable to human activities) (Fig. 3.3).

It should be noted that, while the inorganic component is ever-present, the other two components may not actually be present in the soil under examination, but frequently they can be extremely qualifying if the samples originate from an area, of limited extension, with quasi-homogeneous soil and/or geological characteristics.

3.1.1   The Inorganic Component

The inorganic component of a soil, generally the most abundant, is represented by fragments of rocks and minerals. The identification of the inorganic component can provide useful information on the formation and evolution process of the soil.

The nature of this component depends on the type of rock below the soil, on particular alteration processes operating on it, on the contributions of the materials transported and deposited by the action of water or wind, and on subsequent chemical-physical transformations that take place within of the soil. The formation and the presence of minerals and rocks, as well as their distribution and association in soils are, therefore, related to specific geological and environmental processes, which may also be exclusive in limited areas or extensions under certain conditions, only in specific places. For this reason, the minerals and rock fragments present in a soil can be powerful markers for forensic investigations.

Fig. 3.3 Soil particles observed under stereoscopic microscope. Inorganic particles are visible as well as plant components and red fibers (anthropogenic material)

Generally, the soil scientists divide the minerals present in a soil in two large groups: the primary minerals, i.e. those arising from the chemical and physical processes of disintegration of the rocks (mainly silicates, such as quartz and, to a lesser extent, micas, amphiboles, pyroxenes, and feldspars), and the secondary minerals, of recent formation, known as clay minerals, having dimensions of less than 4 μm.

The primary minerals are, in turn, divided into light and heavy minerals, having a specific weight lower and higher, respectively, of 2.89 g/cm^3. The identification of minerals is essential when it comes to comparing soil samples for forensic purposes (Fig. 3.4).

The primary minerals in soils are often altered. The alteration process includes a set of phenomena which lead to a selective dissolution of the minerals and to a consequent transformation.

The genesis of clay minerals occurs during the process of formation of the soil; their composition thus derives from the characteristics of soil formation, and reflects the environment and history of the soil. Clay minerals are present in the form of very small crystals. The microscope has insufficient magnification power for proper analysis, therefore to identify these crystals it is necessary to use other analytical techniques, including x-ray diffraction, infrared absorption spectrometry or thermo-differential analysis.

The interest in the analysis of clay minerals applied to forensic investigations has been confirmed by numerous studies that have shown that the component of clay minerals can vary significantly according to proximity and topography.

The rock fragments that may be present in a soil are numerous and of different genesis, originating from the alteration of the local rocks and, to a lesser degree, represent contributions of new material by the action of water or wind. However, individual samples and specific groups of them suffer considerable variation from place to place, providing useful information on the geological characteristics of the areas that yield soil samples under examination.

Rock particles are identified by the use of the polarizing microscope, through which certain particular features are identified, such as structure and texture, the

Fig. 3.4 Fragments of minerals present in the soil samples observed under a petrographic microscope with crossed polarizers trim: (**a**) sanidine; (**b**) microcline; (**c**) pyroxene; (**d**) plagioclase

mineral components and their profusion, the amorphous elements, such as glass, solids or fluid inclusions (Fig. 3.5).

The rock fragments in the soils are very small and identifying them is a complex operation; thus, in this analytical phase, the experience and skill of the operator is very important.

3.1.2   The Organic Component

The organic component consists of plants and animal organisms, both living and dead in various stage of decomposition or fossilization, and the complex of humus substances. Humus is a complex mixture of amorphous organic substances with colloidal behaviour which impart to the soil the characteristic brown color.

The plant component includes fragments of leaves, bark, seeds, inflorescences, algae, pollen, fungi, spores.

The surface layer of the soil may contain phytoliths, biogenetic silica particles, formed by the progressive silicification of plant cells. Since the phytoliths take the form of plant cells, each species produces a form of phytolith that is specific to it.

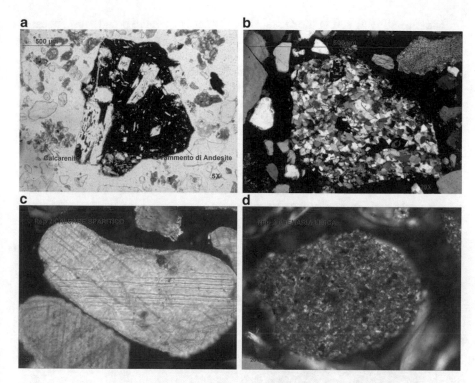

Fig. 3.5 Rock fragments present in soil samples observed under a petrographic microscope: (**a**) andesite; (**b**) sandstone; (**c**) sparitic limestone; (**d**) lithic sandstone

In frequently flooded land it is normal to find a large spread of diatoms, unicellular siliceous algae that colonize all aquatic environments, freshwater, transitional and marine, with a large number of genera and species. Diatoms are an indication useful in the comparison of soil samples for forensic purposes; moreover, their presence in the lungs, liver or kidneys of corpses helps to diagnose death by drowning.

Finally, the study of pollens present in the soil, in addition to being very useful for circumscribing the area of origin of a sample, in which the species of plants referable to the pollen samples are well represented, may provide an effective method of investigation, particularly in those cases where it is necessary to establish the timing of a deposition of a certain soil on a particular area. In fact, the identification and quantity of the pollen species may indicate the season of inflorescence and vegetative propagation.

Animal remains include fragments of marine and land shells, insects, or parts thereof, and microorganisms such as bacteria, protozoa, protists and metazoans. The study of microorganisms in the soil has experienced a great development in recent years, finding interesting applications in the forensic field. In particular, for comparative purposes, some forensic laboratories perform analysis on the DNA of bacteria and other single-celled organisms in the soil.

Although the study of insects in the context of criminal investigation plays a primary role in taphonomic studies, their presence or evidence of their biological activity in a soil can provide useful information on the origin of a sample.

Microfossils fall within the organic component only in the broad sense, because although they were originally organic elements, their fossilization has made them inorganic materials.

The amount and nature of the organic fraction is indicative of environmental features that have presided over the formation and evolution of the soil, with reference to both regional and micro-environment factors. Therefore, the identification and classification of species of plant and animal organisms in the soil may be an additional and useful method for comparing the samples and a valuable tool to determine their origin.

3.1.3   The Anthropogenic Component

The anthropogenic component includes all those fragments or traces of materials and substances, whose presence in a soil is attributable to human activity.

This component consists of fragments of various materials such as paper, plastic, glass, paints, fibers, metals, ceramics, and chemicals, including precipitates, solvents and slag (Fig. 3.6).

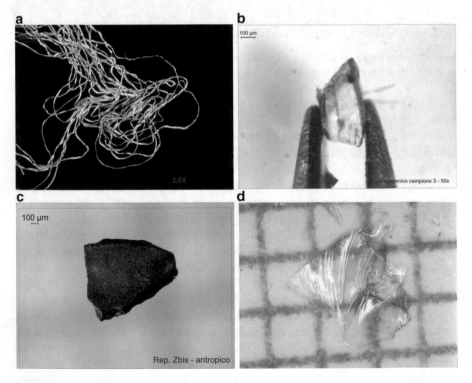

Fig. 3.6 Fragments of anthropogenic materials found in soil samples: (**a**) fibres; (**b**) layers of paints; (**c**) tile; (**d**) yellow glass

Also, there is a considerable variety of man-made materials that, in a sense, can be regarded as artificial rocks; these are the building materials that include cement, mortar, plaster, bricks, and tiles. Some of these materials show textural characters similar to those of natural rocks and in some cases, it is necessary to carry out detailed chemical, mineralogical and petrographic analysis to identify their true nature.

The characterization of the anthropogenic fraction includes a series of analysis whose applicability varies depending on the nature of the component to be examined. The tests which take place on anthropogenic fragments are mostly the morphological observation of detail and physical-chemical analysis to identify the type of material of which they are composed. Such techniques allow not only to identify the material, but also to understand their origin, their function, and their presence in the soil under examination. The presence of such materials in soils can provide strong clues to the extent of being able to differentiate soil samples otherwise similar to each other, or to circumscribe a micro-environment to a zone with similar geological and soil characteristics.

3.2 Interdisciplinarity in Studies and Forensic Soil Science Analysis

As previously discussed in the preceding paragraphs, the soil is a complex system composed of many components of diverse origin and nature. In this regard, to obtain a more complete characterization of the soil what emerges is the need for a detailed analytical procedure for each component present in the soil sample, which inevitably involves the interdisciplinary approach of specific areas of scientific competence.

Before analyzing the detail of the terrain, the forensic soil scientist must separate the components of the organic and anthropogenic fractions that are considered unique and characteristic, to allow them to be studied in greater detail. These elements must then be analyzed and identified by experts of all disciplines relating to the component in question. The experts that are typically called upon are entomologists, chemists, botanists, as well as specialists in materials science. The soil may also contain ultramicroscopic organic components, such as fungi, pollen, bacteria and other microorganisms only visible with microscopic techniques at a very high magnification; their study and analysis requires the services of mycologists, microbiologists and palynologists. Sometimes an area may host human biological elements such as hairs, nail fragments, bone fragments, teeth. Such components must be analyzed by the forensic biologist or by the coroner.

The interdisciplinary study of the soil for forensic purposes is now normal practice adopted in most forensic laboratories. In fact, any element present in the soil, depending on the context in question, can be a good marker to determine the level of comparability between soil samples or to identify a likely micro-environment of origin.

Case Report The usefulness of the analysis of anthropogenic materials present in soil samples is illustrated by an exemplary practical application in the case of damage to a car body, perpetrated by a pair of extortion criminals in Emilia Romagna.

Earlier in the morning, the individuals went to the exercise of a car body repair premises and performed acts of vandalism inside and outside the workshop. In earlier days, the owner had suffered demands for money in exchange for protection, so the investigators were inspired to devote their initial investigations to an extortion hypothesis. Enquiries led to the identification of two suspects and the seizure of their footwear that presented conspicuous soil traces.

The purpose of the soil analysis was to make a comparison between the traces on the suspects' footwear and those found in the soil on the site where the vandalism had occurred. If the traces of soil deposited on the shoes had shown strong comparability with the soil present outside the premises, it would be possible to link the suspects to the crime scene.

The laboratory tests confirmed that the soil traces found on the footwear under investigation showed a strong comparability with the soil of the terrain of the crime scene. All samples showed the same color, similar particle size distribution (that is to say the size distribution of the dimensions of the soil particles) and contained the same mineral species and rock fragments, with extremely compatible relative profusions. In addition, soil samples showed several fragments of singular anthropogenic materials such as transparent micro-spheres, blue colored material fragments, and white material fragments embedding hyaline microspheres which were analyzed by means of targeted chemical-physical techniques (Fig. 3.7).

The transparent micro-spheres had sizes between 250 and 500 µm. The Raman spectroscopy analysis confirmed that the small spheres were made of ordinary glass. The presence of glassy spheres could have conceivably been attributable to *shot peening* operations, i.e. the polishing or non-abrasive cleaning of delicate surfaces such as gears, mechanical parts and metal surfaces. This process consists of blasting, under controlled pressure, the glass micro-spheres on the surface to be treated; the size of the micro-spheres may range from 1 to 800 µm, depending on the area of interest and the required finish. Peening is routinely performed in the automotive industry and in mechanical engineering, aeronautics and electronics.

The particular blue fragments had dimensions variable between 100 and 500 µm. The Raman spectroscopy analysis showed that these fragments contained pigments used for high strength and performance paints and coatings, typically used in the automotive industry.

The white material fragments embedding small transparent spheres had dimensions varying between 50 and 100 µm. Raman spectroscopy revealed that the white material to be of anatase and rutile, consisting of titanium oxide, used for the production of paints. Paints embedding tiny glassy microspheres, known as A-way paints in the relevant sector terminology; they have retroreflective properties and are generally used for painted road and highway signs.

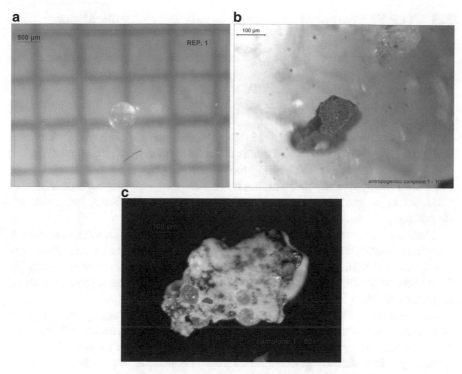

Fig. 3.7 (**a**) Hyaline microsphere; (**b**) fragment of blue colored material; (**c**) white fragment embedding hyaline microspheres

The nature of the three anthropogenic materials and their automotive use indicated an excellent contextual match with the car body repair activity conducted in the vandalized premises. Such materials, in fact, precisely because of their specific use, could not be ubiquitous and are likely to be found only in an environment associated with specific activities such as those carried out in the damaged workshop. Although the gross soil characteristics were less helpful in identifying a specific link between the footwear and the soil of the yard, the anthropogenic fragments added a high level of specificity. The strong comparability shown between the soil samples was confirmed and strengthened by the presence and type of particular anthropogenic elements. The information obtained from geological analysis, evaluated and interpreted in the specific investigative context, assisted the investigators to prove the link between the suspects and the site of the damage.

3.3   The Site Survey in Forensic Soil Science Applications

The aim of the site survey is the collection soil-based evidence and involves different environments, generally outdoor areas, and things, for example motor vehicles, and requires specific methods that have an operational logical sequence, to properly contextualize the specific crime narrative in terms of observation, clue trace searching, and subsequent collection of the samples. In this context, the term trace refers to the soil deposited on a surface during the execution of a crime, while the term sample defines the portion of soil that is collected in situ during the site survey.

It is important to note that during the inspection phases, as well as during the evidence collection of the soil samples, it is essential to wear disposable covering in order not to pollute the crime scene and the traces: overalls, boots, hat, mask and disposable gloves.

During an outdoor inspection, the preliminary stages are those of observation and description of the area, in which the geological, geographic and geometric characteristics are highlighted as well as all the elements that can have a useful meaning in the context of crime and its narrative. The essential actions that take place in these steps are described in Table 3.1.

Through the study of these elements, in addition to the necessary knowledge of what happened at the crime scene and how the events occurred, useful soil traces are searched for, left by those involved in the crime, or the most area suitable for soil sampling is identified by performing an *ad-hoc* collection, based on the environment, the specific crime and its narrative.

Table 3.1 Main actions to be undertaken in the preliminary phase of the inspection of areas and outdoor environments

Action	Description
Preliminary study	It is carried out with support of remote instruments (aerial or satellite imaging etc.) and cartography
Demarcation of the area of interest	It is performed by placing and defining the boundaries, virtual or real
Description of physical and natural features	Highlighting the geological, geographic and geometric characteristics of the location
Description of the morphology	If an area is flat, steep or hilly
Description of anthropogenic factors	For example, embankments, cuts or road embankments, presence of buildings and human activities, etc.
Hydrography description	For example, ponds, canals, runoff, etc.
Description of access routes	Study of access routes to the survey area and their geometry
Description of the vegetation	Study of the type of vegetation and any variation within and outside of the study area
Terrain description	Observation of the color and structure of the soil and any changes of these features inside and outside of the study area
Construction of a topographic map	Plan of area under investigation on which are recorded data of interest: soil sample collection points, type of vegetation present, location of victim discovery or traces of various kinds

3.4   The Collection of Evidence and Preservation of Traces of Soil Samples

The traces and soil samples are sources of evidence and, like all other types of evidence, they must be collected by a specialist in forensic techniques and stored in a suitable way, respecting police procedures. The value of comparisons between soil samples depends largely on how the data was collected and preserved, and the preliminary assessments made in the inspection phase. For this reason, the evidence collection of the soil samples requires a logical sequence of operations, recorded by video or photos documentation at every stage and which are listed, in general, in Table 3.2.

Table 3.2 Main actions to be undertaken during the collection of evidence samples and soil traces

Actions
Suitable disposable covering must be worn; in particular, it is advisable to wear two pairs of disposable gloves
Divide the area of interest into sub-areas according to the different types of soil present, so that each area is represented by a basically homogeneous type of soil
If the area is extended, or has a particular geometry (for example, a narrow and long unpaved road), divide the area into sectors of an appropriate size (of the order of a few square meters) to be assessed depending on the specific case, by using, for example, the presumed crime narrative or the accessibility of the area by a customer or the presence of significant elements, such as fingerprints, dragging tracks, weapons etc.
Within each sector, a single soil sample is taken at the center and at various points around it, then unifying all in a single aggregate representative sample of the sector. Exclusive one-point is not recommended: if sampling is aimed at collecting material to compare with traces that have transferred to mobile surfaces, for example on footwear or tires, it must be considered that these surfaces are in contact with the ground in several places, then the trace it may carry is an average sample of the area on which it transited
Indicate, with a letter or a number, the point from which the soil sample was taken, it must be geo-referenced and localised it with metric references with respect to fixed points and executing a photographic survey
Collect samples using special disposable instruments, such as tweezers, spatulas, brushes; alternately, using steel instruments that between samplings must be cleaned with acetone
Changing the disposable gloves between samplings, in order not to pollute the subsequent sampling activity with foreign material
Let the damp samples dry in the air before placing them in appropriate containers; otherwise, the biological activity may cause the deterioration of the organic components and/or the formation of new ones
Place the samples in containers such as vials, sterile Teflon bags or boxes, in order to ensure the highest possible integrity. Use of only paper containers may allow the passage of external moisture which would lead to the putrefaction of the organic component
Indicate the samples collected with the same code reference used to identify their point of gathering and must be documented photographically
The sufficient amount of sample for the analysis depends on the type of required analysis; in general, a quantity equal to a volume of about 50 ml is required
In the case in which at the crime scene soil traces are found on removable surfaces, such as floor mats and pedals of motor vehicles, footwear etc., collect the entire item and then extract the soil sample in the laboratory

When collecting soil samples in order to compare them with the traces relevant to the crime committed, the number of samples can be very variable and depends on many factors: the extent and geometry of the area of interest, the variability of the characteristics of the terrain within the area, as well as the assumed crime narrative. Generally, it is recommended to collect also samples in areas outside and neighboring the area of interest, to study the variability of soil characteristics in a macro-area and to properly circumscribe a relevant pedological micro-environment.

If items of evidence are seizured from a suspect, these must be inspected at appropriate places, possibly, also carrying out the collection of traces on them (Fig. 3.8).

During the collection of evidence from mobile surfaces, it is very important not to disperse the traces and, in the event there is little material, if possible, it should be concentrated. For example, traces of soil from shoes are usually taken separately from the right and left soles to obtain two samples; in the case of limited quantity of material, the two traces may be combined into a single sample.

Once the collection of samples and soil traces has been completed, the chain of custody of the findings must be assured until the end of the technical examinations, when they are returned to the Judicial Authorities.

Fig. 3.8 Collection, in a controlled environment, of soil sample from a shoe belonging to a suspect

3.5 The Effects of Wrong Methods and Operations During Site Surveys and Evidence Collection

The effectiveness of soil analysis may depend on the way the samples were collected as well as all those activities that are conducted, from the seizure and storage of objects related to the crime, or belonging to a suspect, to the collection of soil trace evidence from the suspect. There is no doubt that the way in which an object was collected and preserved and that the trace sampling method can affect the integrity of the samples and, therefore, on the laboratory analysis. In addition, during the investigative phases, there are critical moments when the underestimation of the potential of analytical methods and the lack of knowledge of their applications can seriously damage the value of achievable results.

In practice, during the stages related to site survey, evidence collection and storage of samples, there may be significant dispersal of material, alteration of traces, the collection of unnecessary samples, and omission of useful trace samples.

The dispersion of significant material and alteration of the traces can take place before, during and after the collection of evidence samples, due to an underestimation of the application of the soil analysis or lack of proper preservation of the sources of evidence. For example, in the case where a motor vehicle is involved in a crime, it would be appropriate to transfer it to a closed environment with a tow truck so as to allow for adequate conservation of the soil traces for a possible later collection of evidence form it.

The alteration of the traces can take place also when the evidence collection is not done properly. There are areas where, day after day, the soil settles, forming a stratification over time; this happens inside automobile fenders. In these cases, if it should be necessary to extract a sample, it is important to remove only the superficial layer, to avoid altering the sample with materials that might not be connected with the crime. A further case of alteration might occur if the collected sample has not been conserved properly. For example, placing a moist soil sample in a plastic container would cause it to get mouldy, adversely affecting the organic component.

The collection of unnecessary samples and the failure to collect useful sampling might occur if a proper assessment is not conducted prior to the survey and collection of evidence. For example, in the case of evidence collection of soil samples from a car, it could be insufficient to collect samples simply from the floor mats, rather extending than extending the search to other surfaces inside and outside the car.

In order to collect soil samples useful to the investigation, appropriate inspection and evidence collection techniques should always be applied to correctly assess and interpret the scene of the crime and the nature of the involved surfaces. In this regard, due regard should be given to the survey procedures and evidence collection methods compiled and tested by forensic institutes and forensic science experts.

3.6   Issues Related to Traces of Soil

The most frequent questions posed to the forensic geologist are: "Is the investigated soil/sediment similar to that from the crime scene? If so, what is the level of comparability?" It must be noted, however, that such trace evidence can be modified during its transfer to an object, and after its deposition. Comparator (reference) sample soils can also change during collection and subsequent storage. Regarding the transferred trace, the inherent nature of the soil/sediment, meteorological conditions during collection, and the force and direction of energy involved in transfer and deposition, are all factors to be considered when processing and interpreting the analytical data.

These conditions may give rise to the situation where surfaces may contain soil traces with characteristics different to those of the place of origin.

The removal, relocation and deposition of loose material are dependent on many factors such as environmental and weather conditions, physical and sedimentological characteristics of the substrate from which the particles are moved, the intensity and the direction of energy that moves, transfers and deposits the particles, and the physical characteristics of the surface on which they are deposited (Fig. 3.9a). After the deposition of soil particles on a surface, considerable changes can affect the trace itself, with dispersion of material and/or the addition of new materials (Fig. 3.9b).

The multivariate nature of geological and pedological evidence traces makes it impossible to establish a useful database. Every case is unique so it is imperative that the trace evidence from each one is studied individually without prejudice. *Ad*

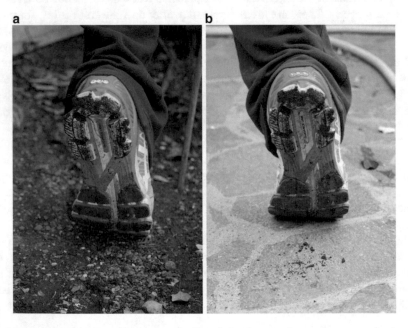

Fig. 3.9 Removal and deposition stages of soil particles from a substrate by means of a shoe (**a**) and subsequent dispersion of due to the movement of the material (**b**)

hoc studies could be carried out in attempts to strengthen the probative value of soil evidence. Additional information gained from such studies might fill gaps in the analytical data. The *ad hoc* study should be executed jointly with the investigators. It is important for the analyst to know (i) weather conditions before, during, and after the criminal offence, (ii) the likely narrative of the criminal act, (iii) the probable narrative after the crime, and (iv) the lag between the time of the offence and seizure of exhibits.

When possible, it is necessary that the collection of samples and traces of land is conducted by a geologist, experienced in forensic techniques, who knows what, where and how to perform the collection and who is fully informed about the way in which the crime was committed and on the related narrative; not only to facilitate the investigation of the crime scene and the collection of representative samples, but also to understand how and to what extent the samples deposited on various surfaces can be comparable with those sampled on the site of a crime.

The most common effects that may occur, even simultaneously, during the removal, relocation and deposition of soils and the subsequent modifications of the same deposit are selective deposition, dispersion and mixing.

3.6.1 Selective Deposition

Selective deposition occurs when not all the soil components are transferred onto a surface, whether they are inorganic, organic or anthropogenic in nature. Surfaces and fabrics can vary in their ability to retain trace evidence, so transference will involve an element of selectivity. This can be a function of clasts size and mass. For instance, only fine material, such as silt and clay, might be deposited on a shoe when it has been in contact with dry soil, and coarser particles might be left behind. In this case, only the finer fraction of the soil sample will be comparable with those of the comparator sample. Another case might be that in which the gravel granules are trapped in the groove of the sole of a shoe or in the spaces of the tread of a tire, while the finer particles remain in place because they remain aggregated in a state of apparent cohesion.

It is clear that the selective deposition of the granules is strongly influenced by the moisture content of the soil, the weather conditions, the energy and the type of surface that mobilize the particles.

3.6.2 Dispersion

Dispersion, or loss of, particles after transference also affects the level of comparability possible with a reference sample. For example, losses from the wheel of a vehicle are correlated with the distance of journeys after the transfer. The speed of wheel revolution is also pertinent. By way of trial, a test carried out in Italy showed that the soil present on a car wheel after a three- mile drive was 70% less than at the start. In this case, the loss of soil components meant that the soil trace was certainly not representative of the original soil that was transferred to the wheel. There are also cases in which the material deposited on a surface is dispersed before being sampled for environmental or weather causes, for example to track exposure to strong rain or wind, or for faunal activities that interact with different mode with the scene of the crime.

If leakage occurs, even partially, of traces of soil, the low-quantity sample material may not be representative of the origin terrain, and results are lacking in several components. In some cases, it is possible to establish that trace dispersal has occurred; in others, there may be a reasonable suspicion of dispersal but without total certainty. In both cases, during the processing of the results, the soil samples may not have significant and sufficient comparability to determine whether they came from the same micro-environment or they may show only partial information on the likely places of origin.

3.6.3   Mixing

Mixing takes place when on a given area are deposited at different times, also chronologically distant from the crime.

Mixing is important to the consideration of the temporal aspects of deposition. On an item such as a shoe, or a foot well mat of a vehicle, soil can accumulate both before and after the soil from the crime scene has been transferred. With repeated use, and absence of cleaning, there is sequential deposition into a car foot well from footwear. The sequentially-deposited material can be considerably different, or be comparable, to the reference sample. Thus, it may be difficult to distinguish even the most recent deposition from the original material accumulated at the time of an offence. This imposes obvious limitation on interpretation of mineralogical data.

Very few mineralogical species can reflect season or weather. However, because of the predictability of flowering and sporing times, analysis of plant macro-remains, plant spores, pollen and fungal spores, might give more information on temporal aspects of deposition. Even here, there are difficulties because of the residuality of pollen and spores. They can remain on a surface for considerable periods (many years under suitable conditions) so that seasonal effects can be confused. Nevertheless, under certain circumstances, botany and mycology can provide valuable temporal evidence.

Case Report The case illustrated below is a good example of how some measures taken during the inspection and evidence collection of traces can be useful to discriminate the different types of soil deposited at different times on the same surface.

In 2003, in the province of Catanzaro, a convicted felon on probation was killed on his way to the Police Station for the daily signature. Witnesses said that the killer, who was travelling alone on a motorcycle, had approached the man and then fired several gunshots. The day after the murder the police found the motorcycle described by witnesses in open country, lying on the ground. The investigation converged on a suspect whose car was promptly seized. The investigators hypothesized a crime narrative: the killer, after killing the man, had abandoned the motorcycle in the countryside, where he had previously left his car, and then abandoned the area with his own car and dispersed his tracks.

The suspect's car was inspected to seek clue traces that could corroborate the narrative assumed by the investigators and thus link the suspect to the murder. Front driver and passenger floor mats presented soil deposits and although the witnesses had seen only one man on the motorcycle and the investigations had highlighted that no other persons were involved in the crime narrative, the traces from both mats were collected. Evidence samples were also collected from the terrain where the motorcycle had been found to compare these samples with the traces from the car. If the samples had presented a strong comparability, it would have been possible to connect the suspect's car with the place where the motorcycle had been abandoned.

From careful observation and analysis of the traces it was established that the deposit on the driver's side mat was a mixture of at least two different types of soils, evidently deposited at different times. In particular, the deposit was characterized by the presence of different types of particles: grains with angular morphology comprising quartz, kaolinite and plagioclase minerals as well as fragments of rock such as limestone and metamorphites; grains with rounded morphology consisting of the quartz, calcite and aragonite minerals and fragments of rocks such as limestone and sandstone. The first type of particles had very similar characteristics to those of the soil samples taken at the place of discovery of the motorcycle. The second type of particle most likely come from a marine coastal environment, however, there were no certain indication of which trace particles were unrelated to the criminal narrative, therefore the sample from the driver's side mat was not comparable.

The soil trace taken from the passenger side floor mat contained particles very similar to the second type of grains described above and it was therefore reasonable to assume that both had been deposited in the same context. This evidence allowed for subtraction from the sample found on the driver's side floor mat all the elements analogous to those found on the passenger side floor mat. The result was the identification of a soil typology extremely comparable to that found at the location where the motorcycle had been found, in terms of color, morphology, as well as mineralogical and petrographic characteristics.

In this case, a trace of soil unrelated to the crime narrative assisted the investigators in hypothesising a certain succession of events leading to the placing of the suspect's car at the spot where the motorcycle had been abandoned.

Chapter 4
Forensic Analysis of Soils and Geological Materials

Rosa Maria Di Maggio

Abstract The characterization of geological materials envisages, in general, the use of a few analytical techniques targeted to the kind of result to be achieved. The characterization of soils, however, requires a greater number of targeted analysis to obtain suitable information for the comparison of multiple samples, and for the determination of a likely area of provenance. The examination of soils must be carried out following *ad hoc* analytical procedures which have a logical sequence and that are able to better characterize the samples on the basis of their distinctive elements.

Keywords Soil analysis • Microscopy • X-ray • Spectroscopy

The characterization of geological materials envisages, in general, the use of a few analytical techniques targeted to the kind of result to be achieved. The characterization of soils, however, given their heterogeneous composition, requires a greater number of targeted analysis to obtain information as possible, is suitable for the comparison of multiple samples, and for the determination of a likely area of origin.

Generally, the soils involved in forensic analysis are samples which have been handled. Such handling tends to obliterate or irreversibly alter the purely chemical properties of a soil, such as pH, redox potential, or the cation exchange capacity, and some physical properties, such as temperature or porosity, which have characterizing value only if analyzed in situ or on undisturbed samples. The analysis of the soils is, therefore, mostly dedicated to the physical-chemical characterization of the various types of particles that compose them and to the study of physical characteristics that are not affected by handling, such as color. The examination of soils must be carried out following *ad hoc* analytical procedures which have a logical sequence and that are able to better characterize the samples on the basis of their distinctive elements.

R.M. Di Maggio (✉)
Geoscienze Forensi Italia® – Forensic Geoscience Italy, Rome, Italy
e-mail: dimaggio@geologiaforense.com

© Springer International Publishing AG 2017
R.M. Di Maggio, P.M. Barone (eds.), *Geoscientists at Crime Scenes*,
Soil Forensics, DOI 10.1007/978-3-319-58048-7_4

4.1   ForensicLaboratory Analytical Procedures for Soils

Although forensic soil analysis dates back to the first decades of the last century, only in recent decades forensic geologists have begun to establish analytical laboratory procedures in a guideline framework. Although we do not yet have a standard reference universally accepted procedure, forensic laboratories specialized in soil analysis implement their procedures based on a logical analytical sequence and the optimization of techniques to get results that are unique as possible and representative of the specific characteristics of the examined samples.

In general, all procedures include preliminary analytical techniques that provide general guidance on the characteristics of the soil sample, and detailed analytical phases (the choice of which varies according to the particular characteristics of the particles of the substances and materials in the soil and the information that can be obtained) that can give more detailed information on the individual particles that compose it. In general, it is necessary to implement diverse analytical techniques including colorimetric, sedimentological, morphological, mineralogical, petrographic, crystallographic, micropaleontological, chemical, physical, and biological approaches.

The optimum progression of the analytical techniques depends on a number of factors such as the amount of available sample and the results obtained in the earliest analytical steps. In this context, the results of the preliminary examinations suggest the most appropriate analysis to be performed in the subsequent procedural steps.

Soil analysis can provide two types of information, depending on the specific context, the crime narrative and the requirements of the investigation:

(i) the level of comparability between multiple samples and/or traces;
(ii) the location of the possible geographic origin of a trace.

While in the first case it is necessary to have available two or more samples, in the second case, the information can also be sought with a single specimen. In both cases, the traces should have a sufficient amount of material; indeed, a low amount may limit the number and the validity of the laboratory analyzes with the risk of not getting sufficient data to perform comparisons or to determine a possible location of origin. What is of particular importance, however, is not so much the weight or volume of the material to be analyzed, but the availability of a sufficiently large and statistically significant number of particles. It is also essential that the amount of the sample is adequate to ensure that it is representative and to allow the execution of analysis of all types, including destructive ones. It is therefore possible to operate with minimal quantities with the availability of a high number of components capable of identifying discrete associations of particular characteristics that help to determine strong comparability with comparison references. In practice, in several tens of milligrams of material it is possible to distinguish the nature of the ingredients and their reciprocal composition ratios.

The level of comparability between soil samples is estimated by means of the comparison of geological, physical and chemical (qualitative and quantitative) data obtained in the analytical phase and used for terms of comparison.

The localization of an area through the characterization of soils, and more generally, of geological materials, can be carried out effectively because the relevant geological, pedological, mineralogical and petrographic characteristics can be extremely distinctive of a given micro-environment and it can vary over short distances. Generally, the trace data obtained in the analytical phase is interpreted to determine the type of soil and/or geological environment of the origin location. The acquired information is interpreted through cartographic studies (geological maps, soil maps and so on) in order to determine a probable geographic origin location. At this stage, it is very important to correctly read and interpret the available maps with the support of investigative information. Having identified a possible area of origin, relevant samples are collected to make comparisons with available traces so to confirm or exclude its validity.

The study and evaluation of the level of comparability between soil samples and/ or of their likely geographical origin must necessarily be integrated with a preparatory study on the variation of soil characteristics within a macro-area. In fact, only by circumscribing micro-environments in which the soil geological and environmental characteristics are almost homogeneous is it possible to get more precise and professionally correct answers regarding the probable origin of the traces or on their level of comparability. In addition, the forensic geologist, as well as performing a thorough analysis of the evidence, must process and interpret the results and place them in the specific context of the actual or presumed crime narrative. We must not overlook the fact that superficiality and/or inadequate expertise reflect negatively on the probative validity of the evidence and the outcome of the trial.

The development of forensic soil science that has occurred in recent years, thanks to the organization of numerous international conferences and workshops, has spurred many experts to implement research and experimentation and to improve technical and analytical procedures. For this purpose, different projects were launched by international groups engaged in forensic geology, notable among which is the Initiative on Forensic Geology working group (IFG), as part of the International Union of Geological Science (IUGS), that, in addition to pressing for a greater development of all the applications of forensic geoscience, is preparing the publication of guidelines for the study and analysis of soil that can be considered as a useful reference by experts working in this field.

The following are some of the most commonly adopted techniques for the determination of soil samples. For further analysis and technical details, please refer to specific texts.

4.2   Color Analysis

Since ancient times, color has always been an important diagnostic criterion for the recognition and evaluation of soils as it provides an indication on their formation environment, constituting one of the most evident characteristics.

The color of soil is the result of a set of factors; among the most important of which are the amount and types of organic substances, the nature of the minerals present and the oxidation and reduction of iron. Initially, the color of soil was attributed in a very subjective way using qualitative terms, without the application of any conventional methodology. Since 1954, all the soil colors are determined by comparison using the Munsell system, which involves the use of colorimetric tables specifically created for the study of soil colors.

In the field of forensic pedology, as well as for soils, the determination of color is an analysis applied to all kinds of fine-grained loose material. To classify soil, generally the Munsell color system is used, but the colors can also be measured instrumentally by using the latest generation spectrophotometers. Color is therefore a useful feature in soil interpretation that, along with others, allows for the recognition and demarcation of representative units of terrain in a specific area.

Among the soils, over 1100 different colors can be distinguished; this wide variability ensures that color is a potentially useful method for discriminating the soil samples.

The Munsell Soil Color Charts is a manual that includes a series of tables, each containing about 25 colored tiles, particularly relevant for the determination of the soil color. Between the tiles there is a hole where the soil sample to be classified is placed.

Each color is described with a standard qualitative term, and classified according to an alphanumeric combination of three parameters: hue (H), value (V) and chroma (C). The color coding is expressed in the form H V/C. Hue indicates the tonality of the color and is divided into five basic components (red, yellow, green, blue and purple) and physically indicates the relative amount of each component in a given color. In the Munsell system, the hue value is identified by each table containing the colored tiles. The value is the level of brightness or darkness of a color, while the chroma, or purity, defines the saturation. Within each table, the value and chroma vary respectively vertically and horizontally.

Generally, the forensic pedologists prefer a color classification based only on the clay fraction in soil samples, to minimize the variations due to their heterogeneity. Indeed, it is precisely clay that envelops the mineral particles and gives color to the soil.

To obtain a homogeneous clay sample from a sample of rough ground, a suspension of the sample is filtered through a sieve having a mesh with a size of 4 micrometers so as to separate the clay. Subsequently, the clay fraction is treated in hydrogen peroxide, to decompose the organic materials contained in it, then it is dried and homogenized.

4.3   Particle Size Analysis

The mineral elements and the rock fragments present in soil show, in general, very variable dimensions. The particle size, also called texture, is the measure of the particle size of a sediment. The erosive processes, transport and sedimentation greatly influence the size of the granules of a clastic deposit, as also those of a soil.

In the classification by size of clasts according Wenthworth, the particles are classified according to their diameter into size classes called gravel (with particles having a diameter greater than 2 mm), sand (with particles having a diameter of between 0.0625 and 2 mm), silt (particles with and having a diameter of between 0.0039 and 0.0625 mm), and clay (with particles having a diameter of less than 0.0039 mm). The texture of a soil is defined by the percentage by weight with which gravel, sand, silt and clay are present. Since the size fractions are present in various percentages in the different soils, they will have different textural designations, such as clay soil or gravel-sand, sandy-silt, and so on.

In sedimentology, the analysis of particle size is meaningful only if carried out using statistical criteria; in forensic soil science, this is not always possible because frequently the soil samples collected from various surfaces can be limited in weight or they may not represent the original texture due to dispersion phenomena or selective deposition. However, when the type of the samples allows, graphical representations of particle size distributions are created; also, the statistical parameters are determined to quantitatively analysis the particle size distribution curves and use them as a comparison element between samples.

When the sample is limited in quantity, the texture is determined only in a qualitative manner. It may happen that the soil samples are so limited in weight that the particle size, even if determined on a qualitative basis, would still not be sufficiently indicative to characterize a sample, and to make a comparison with any reference soils.

Before performing the granulometric separation, the soil samples are immersed in water and treated with ultrasound to eliminate the state of apparent cohesion. Subsequently, the loose particles are separated, dry or in water, by means of a sieve tower, with meshes decreasing of standardized dimensions. The sieves are placed on a vibrating machine, the movement of which enables the soil particles to descend by gravity through the mesh and to distribute themselves according to the size.

The percentages by weight of the granules size classes are determined by weighing each fraction which pass through the sieve with the selected mesh size. The ratio of the weight of these fractions to the total sample weight of the multiplied by 100 defines the percentage of particles of diameter less than or equal to that of the sieve mesh. The percentages are used to obtain the characteristic particle size distribution graphs of the examined sample. The graphical representations of particle size distribution can be of various types, such as histograms, cumulative frequency curves or simple or probabilistic cumulative curves; the latter are used for the determination of the statistical parameters that numerically indicate the salient features of the granulometric curves.

Generally, the process of separation by sieving is carried out for decreasing particle size down to sand or silt. However, for several years the market has made available sieves with meshes having a size of 4 μm, by means of which it is possible to separate the soil clay fraction in water. Alternatively, other methods can be used for the separation of the clay fraction; the best known is particle size analysis by sedimentation, or aerometry, which, according to Stokes' law, is based on the physical evidence that a solid particle immersed in the water falls downward with a speed that depends on its size, its specific mass and viscosity, and the specific gravity of the liquid in which it moves. This method consists in dispersing in water the fine part of a soil sample and in quantifying the diameter of the granules by measuring their different precipitation speeds.

4.4 Density Analysis

The density (more accurately called volumetric mass density) of a body is defined as the ratio between the mass of a body and its volume. The inorganic particles of a soil, therefore minerals and rocks, as also all the other geological materials, have known densities that depend on the distribution of their atoms in space.

Forensic laboratories have already been using for about 50 years the division of particles and aggregates as a function of density for comparing soil samples. The process consists of filling a tube of liquids with different densities, from the heavier, on the bottom, the lightest, on top. These reference liquids are mixtures of light liquid and heavy liquid in different proportions: generally, bromoform (2.89 g/ml) or tetrabromoethane (2.96 g/ml) for the heavy liquid, and bromobenzene (1.50 g/ml) or ethanol (0.789 g/ml) for light liquids. The soil samples are dried and homogenized and subsequently inserted in the tube, where the particles are separated from the liquid as a function of their density.

Many scholars have conducted extensive studies on the distribution of soil particles as a function of density and have shown that most of the samples corresponded to a distribution scheme identical to that of other samples taken at other places. Furthermore, the results of the comparisons varied significantly depending on the observer. In the past, experts have tended to recommend the observation of the distribution of particles as a function of the density to compare the soil samples. To date, it has been understood that this test provides very limited answers and can be useful only in specific cases and in association with other characterizations.

The determination of the density of the coherent geological materials, such as precious gems and stones, metals and rocks, is generally performed using the principle of Archimedes, with the method of gauging the mass with immersion in distilled water, by means of a hydrostatic weighing scales. This methodology allows determining the weight and the volume of the sample, values required to calculate density. The procedure consists of the following steps: the sample is weighed in air and the operation is repeated with immersion in water; the difference between the two weights measured corresponds to that of the liquid that has been moved, that is, an amount of water whose volume is equal to that of the test sample. The density then corresponds to the ratio between the weight of the object in air and the weight loss that occurs when it is immersed in water.

4.5 Stereoscopic Microscope Analysis

The stereoscopic microscope, or stereo-microscope, is a tool designed to produce a spatial view of the test sample. To do this, the instrument is provided with two optical paths that are separate differently aligned to provide differently angled images to the left and right eyes. The illumination of the object is carried out by reflected light, which allows for the study of thick or opaque objects; modern stereo-microscopes are, however, also equipped for transmitted light for the observation of transparent three-dimensional objects.

The stereoscopic microscopy allows for repeatable examinations, the study of solid sample surfaces and the examination of details that are not visible to the naked eye or with low magnification. The stereo microscope enables different magnifications, even up to 250×, however, one must always bear in mind that the very high magnifications do not allow for a wide depth of field.

In forensic soil science, stereoscopic observation allows the study of the morphological characteristics of the soil particles, determining their angularity or roundness, brightness, shape and any other noticeable feature (Fig. 4.1); also, it allows for the observation, separation, description and characterization of any organic and anthropogenic components, considered useful for forensic purposes. Generally, observation under the stereoscopic microscope is performed several times during the sequence of analytical stages. The samples are, in fact, first observed as they are, without effecting any preliminary treatment, to observe the main characteristics; subsequently, they are examined after washing in the ultrasonic bath and sieving. In this phase, the observation of each individual sieve fraction is made to study the morphological and morphoscopic characteristics of the individual particles and to separate particular elements on which to perform targeted analysis. The shape and the surface of the soil particles are mainly conditioned by their mineralogical and textural characteristics and the size and shape of the initial fragments; other conditions being equal, determinant factors are the duration of the erosion and transport modes, the mode of sedimentation and the chemical-physical alterations endured after deposition. The shape and appearance of the surface of the particles are, therefore, properties that can give important information on their transport and deposition environment and can be used as a term of comparison between the soil samples. In fact, for equal mineralogical-petrographic components of a given size, the granules of soil samples may present dissimilar surface shapes and appearances if they evolve differently. The description of the surface of the granules contemplates visual observation and the use of one or more qualitative terms; thus, the surface can be rough, glossy, opaque or fractured. The description of the shape provides for angular and linear measurements respectively of the edges and the axes of the granules. Quantitative data is expressed in qualitative terms that describe the degree of rounding, the angularity and the elongation of the particles.

The observation of the surface and the shape of the granules can be carried out with different means of investigation depending on the size of the examined particles. The larger particles are observed with a stereoscopic microscope interfaced

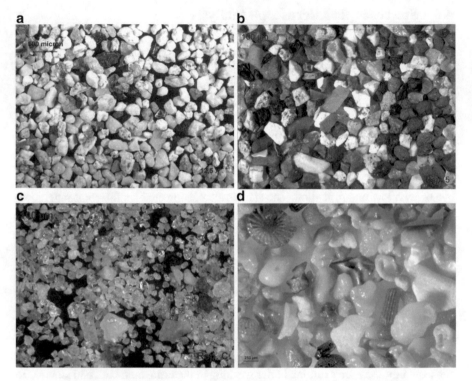

Fig. 4.1 Forensic soils observed with stereoscopic microscope: (**a**) agricultural soil characterized by sub-angular clasts; (**b**) soil from a volcanic area with clasts in sub-rounded and sub-angular morphology; (**c**) beach sand with sub-rounded clasts; (**d**) beach sand with clasts in sub-rounded and rounded morphology

with a software that provides linear and angular measurements; for the very small particles, it is necessary to use a scanning electron microscope.

In forensic geology, observation under the stereoscopic microscope is carried out on all those geological materials for which it is useful to study the morphological, geometric and microscopic characteristics. For example, in the study of gems, dark-field stereo-microscopy allows for an accurate observation of the cuts, the geometry of the faces, the finish and above all, the amount of inclusions and structural distortions that determine the degree of purity.

4.6   Analysis by Means of the Polarizing Microscope

The petrographic microscope with polarized light allows the thin section study and identification of minerals and rocks. In forensic geology, analysis by means of petrographic microscope is useful to carry out an accurate identification of minerals and rock fragments constituting the soil particles (Fig. 4.2), and the study of

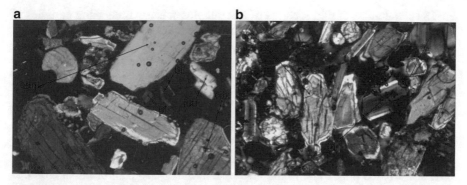

Fig. 4.2 Comparison of soil samples using polarized light and cross-polarization. Sample (**a**) is from the shoe of a suspect, while sample (**b**) is from a volcanic area. Legend: *bi* biotite, *cpx*-clinopyroxene, *leu*-leucite, *pl*-plagioclase, *san*-sanidine

individual geological materials such as minerals, rocks, microfossils and anthropogenic materials such as cement or bricks.

The polarized light microscope consists of two basic parts: the stand, composed of a base and an arm on which are fitted the screws for the focus adjustment, and the optical system, divided into a lower part and an upper part with respect to a rotating plate on which is placed the specimen to be observed. This microscope, in addition to being an instrument that magnifies the image, comes with a light polarization and analysis system and an accessory that allows the optical path of the light to be changed from straight (orthoscopic condition) to conical (conoscopic condition).

Observation consists of illuminating the sample with transmitted polarized light, using different set-ups and conditions, which, crossing the crystal structures, gives rise to a series of optical interferences. The polarized light interferences occur as a function of the crystalline structure of the minerals, so they are able to provide detailed information on the mineral species in question.

The observation under a petrographic microscope is also extremely useful for studying the geometrical and morphologic features of specific minerals and characterizing rock structures and textures.

Polarizing microscope observations of soil sample thin sections also allow for semi quantitative analysis of the sample components. This analysis is performed with the aid of an integrator table which allows micrometric linear displacements to be applied on the sample, according to the two fundamental x, y directions on a Cartesian plane. By means of these movements it is possible to perform a count, manual or electronic, of a number of granules of each mineralogical species (from 1000 to 2000 units). The granules of each species are counted separately, so that there as many partial sums as there are mineral phases observed. Subsequently, the next stage is the calculation of the volumetric percentages with which each mineral is present in the section of terrain. This methodology is appropriately applied when the size of minerals is fairly homogeneous. However, if the section presents minerals of significantly different sizes then some corrective factors can be applied to the calculations.

The preparation of thin sections of soil samples consists immersing particles of grain-size fractions of sand in epoxy resins to obtain coherent and solid cylinders. The cylinders are cut using circular saws, or microtomes, into sheets about 100 µm thick, which are mounted on microscope slides and then thinned and polished manually with abrasive powders, until reaching a thickness of about 20–30 µm.

The determinations in polarized light of a thin section follow a logical operational sequence. The first observations to be made are those in which the direction of propagation of the polarized light is parallel to the axis of the microscope (orthoscopic conditions).

In orthoscopic conditions, there are two possible arrangements of light: polarizer only and crossed polarizers. In the first case, in the optical path of the microscope only the polarizer is inserted and examinations can be carried out on the mineral for aspects concerning the morphological appearance, the measurement of refractive index, the color and pleochroism, the presence of inclusions, the degree of alteration, traces of corrosion, and the type of fracture and flaking. The structure with crossed polarizers involves the use of a second polarizer, called an analyzer, whose vibration plane is at right angles to that of the polarizer. Crossed polarizer observations can analyze all light interference phenomena that occur inside the mineral when the polarized light passing through it is broken down in two vibration directions. By means of this setup, the birefringence, the optical determination of the optical and crystallographic orientations, and the study of anomalous interference colors can be analyzed (Fig. 4.3).

Subsequent determinations are carried out in conoscopic conditions, where the polarized light beams are made convergent by the use of an additional lens. Such alignment of light produces an interference pattern that allows for the study the optical sign of the mineral, as well as the determination of the value of the angle of the optical axes.

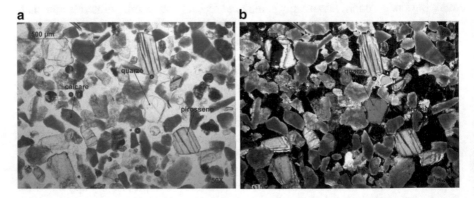

Fig. 4.3 Calcareous soil sample, observed under a petrographic microscope alignment with only the polarizer (**a**) and in crossed polarizers (**b**). Legend: calcare – limestone, quarzo – quartz, pirosseno – pyroxene

4.7   X-ray Diffraction Analysis

The x-ray diffraction (XRD) is an analysis which allows for the recognition of the organic and inorganic compounds having a crystalline structure, or materials that present at the atomic level certain characteristics of order and symmetry.

X-ray diffraction analysis is based on the spread of the radiation x of the material when it is irradiated with a monochromatic beam of x-rays. The x-rays, having wavelengths comparable to the interatomic distances of the crystalline phases, giving rise to the phenomenon of diffraction. Using a crystal as a x-ray diffraction prism, it is possible to determine the shape and size of the elementary cell of the crystal, that is to say how the atoms are distributed spatially.

The most commonly diffractometer used is the powder diffractometer in which a parallel and monochromatic beam of x-rays, a known wavelength, affects the sample, previously pulverized, with different angles. Radiation is diffracted from the sample according to different angles as a function of the characteristic interatomic space of each crystal. Such radiations are intercepted by a detector which transforms them into electrical pulses which are amplified and transmitted to a computer carries out the mathematical and graphic processing. The result of the data processing is a diffraction spectrum, known as a diffractogram. The diffraction spectrum consists of a series of peaks having intensity and angular position relative to the various crystalline phases present in the sample. For the determination of the crystalline phases, the peaks are compared with the reference standard database (Fig. 4.4).

Through x-ray diffractometry, it is also possible to perform a quantitative analysis of the crystalline phases present in a sample using the Rietveld method. However, the proper application of this method requires a calibration with reference samples or knowledge of the crystal structure of the phases present in the sample and depends greatly on the handling of the sample and the orientation of the individual crystals.

The powder diffractometer presents numerous analytical advantages, in fact it allows for the study of the diffraction of x-rays also on materials in the fine aggregate or incoherent states; furthermore, it allows for the characteristics of a crystalline phase to be measured even in the presence of other phases, without there being any significantly interference on the measurement. If quantitative analysis are not required, the substance does not require special care in preparation; the material must be simply pulverized, an operation which is usually performed manually by grinding in an agate mortar. Finally, for the analysis just a small amount of substance suffices and during the analysis it does not undergo any chemical-physical alteration.

In forensic geology, X-ray diffractometry is applied for the identification of all those geological materials that exhibit crystalline structure such as generic minerals, clay minerals and rocks. Diffractometry is also ideal for the study of numerous anthropogenic materials with a crystalline structure which can be present in soils such as cement, bricks, paints.

Case Report

In April 2005, two women, wife and daughter of a convicted felon who was incarcerated at the time of the events, were killed and buried in the grounds of a residence

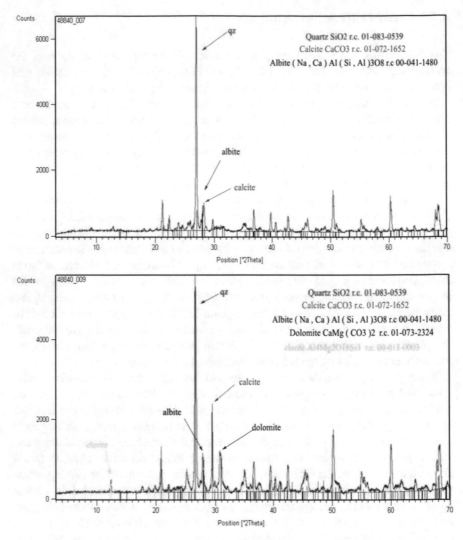

Fig. 4.4 Diffractograms of two compared soil samples. On the abscissa are represented the values of the diffraction angles expressed in 2θ, on the ordinates the values of the intensity of the beam diffracted from the crystals. The two samples have different crystalline phases

in Mirabello, in the province of Campobasso. The investigations made it possible to establish, at the earliest stages, a crime narrative hypothesis, and to identify the culprit and some suspects.

Angelo Izzo, best known as one of the leaders of the Circeo massacre, which took place in 1975, taking advantage of the state of parole with day-release underway since December of 2004, had killed the two women and had concealed their bodies with the help of some accomplices.

The bodies of the two women were buried in a shallow grave and covered with earth and compact white material like hydrated lime. Probably, the bodies had been covered with lime powder, which had been subsequently wet, with the aim of creating a cover to prevent the leakage of decomposition odors (Fig. 4.5a).

To confirm the suspicions of Izzo's likely accomplices, an inspection was carried out on their vehicles in order to discover traces that could connect them to the crime. No particularly interesting tools were found in their cars, such as shovels or spades, or significant soil traces that could indicate excavation activity on the part of the suspects. However, traces of white powdery material were found in the boots of the cars (Fig. 4.5b). It was important to understand if the white material present in the car boots had leaked from lime used later to conceal the victims.

X-ray diffractometry was carried out to determine the composition of the crystalline phases present in the samples. The white powder found in one of the two trunks had the same composition of the material gathered at the pit where the two women were buried: portlandite and calcite. Portlandite is a calcium hydroxide, in building material terminology called hydrated lime, used as a binder for the production of mortars and plasters, usually available in the forms of powder or paste on the market. The sample collected from the pit showed Portlandite in greater amounts compared to the findings inside the car, indicating it had been wet. In fact, when hydrated lime is wet, many crystals of Portlandite are formed, whose magnitude increases with time up to a maximum of three years, thereafter the size of the crystals reduces as the lime ages.

The comparison between the traces of white material found in the suspect's car and the lime collected at the pit, allowed for useful information to be added to the investigation, linking the suspect to the crime narrative. He had actually bought bags of powdered lime and had transported them with his car to the house where the women were killed, thus becoming an accomplice of Izzo in the concealment of the corpses.

Some defective lime bags had left traces in his car which subsequently helped to prove his involvement in the crime.

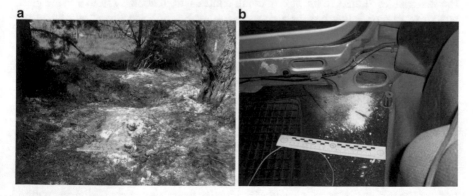

Fig. 4.5 (**a**) The shallow grave and the white hydrated lime; (**b**) the white powdery material found in the boot of car

4.8 SEM-EDX Spectroscopy

Scanning electron microscopy (SEM) and x-ray microanalysis (EDX) is a non-destructive analysis technique for the determination of chemical elements present in a sample and to the study of the morphological details.

The technique consists of irrigating the sample to be analyzed with primary electrons, generated by thermionic effect from the microscope. The primary electrons, before hitting the sample, are accelerated by a potential difference variable between 1 and 30 kV, then collimated and scanned by means of a lens system. The interaction between the primary electrons and the sample produces several phenomena, such as the emission of secondary and back-scattered electrons, which are two of the most commonly used, x-ray, UV and IR radiation signals.

The secondary electrons, suitably collected and processed by a detector, provide information on the appearance of the sample and allow to reconstruct a grayscale three-dimensional image, which is used to obtain morphological and structural information. The number of secondary electrons is a function of the morphological characteristics of the scanned area, the higher the number of emitted secondary electrons, the lighter is the image detail. The back-scattered electrons provide information relating to both the morphology that the composition of the sample. The x-rays of the atoms present in the sample, having energies characteristic of the atoms from which they derive, allow for unique identification of the composition of the observed material: by measuring the energy and distribution of the intensity of the x-rays generated by the electron beam on the sample, using a appropriate detector, non-destructive qualitative and quantitative analysis can be carried out on the chemical elements in the scanned sample.

For the observation of a sample, it is necessary to create a vacuum within the microscope with pressures ranging from 10^{-4} to 750 Pa in order to reduce the interaction between electrons and gas molecules. Modern electron microscopes can have magnifications of the order 150,000–200,000, with a resolving power of the order of one millionth of a millimeter; for this reason the detail and finesse provided by the SEM images, thanks to the strong detail contrast, high depth of field and the great spatial resolution, allowing for a comparison of the different morphologies at the ultrastructural level.

Scanning electron microscopy has become one of the most modern and avant-garde investigative techniques, today finding applications in many areas of science and technology of materials, natural and environmental sciences, as well as in many branches of forensic science. In fact, it is used for the study and analysis of residual particles of firearms shots, ballistic traces, as well as a wide range of commodity, botanical, zoological and geological materials.

In forensic geology, scanning electron microscopy allows us to study in detail minerals, gems, microfossils, rocks, metals, dust and pollutants. For example, in gemology it is used to carry out detailed study of the inclusions and structural deformations of the surfaces of gems, as well as to identify their exact composition and nature. In the study of metals, it is used to identify the elements that constitute the alloys and quantify their exact percentage. In the environmental analysis context, it allows for the identification and quantification of contaminants of various matrices such as air, water, soil and waste. For example, analysis of the solid particles present in the air, such as PM10, allows for the identification of the chemical composition. In forensic soil science, as well as identifying the individual inorganic, organic, and man-made particle elements present in the soil, this technique applies to detailed morphological analysis; a rough appearance on the particle surfaces may indicate their transport and deposition environment, providing useful information both for tracing the probable origin of the soil sample, as well as performing accurate comparisons.

4.9   Raman Spectroscopy

When monochromatic electromagnetic radiation is incident on the surface of an object it can interact with its molecules with three main modes: reflection, absorption and scattering. Scattering involves a minimal part of the incident radiation and occurs because the molecules of the irradiated object, responsible for the scattering, have very small dimensions compared to the wavelength of visible radiation. If the scattering occurs by elastic interaction between matter and photons, then the scattered radiation has the same energy as the incident; if the scattering is a consequence of an inelastic interaction, then the scattered radiation has energy respectively smaller and larger than the incident, and it is called Raman scattering (named after the Indian physicist C. V. Raman, Nobel Prize winner in 1930, who described for the first time the inelastic scattering of photons). The signals corresponding to inelastic interactions with which photons of lower energy are emitted are related to the functional groups of the molecules of the sample and their vibrations, and they are therefore used to identify the compounds present in the sample.

Raman spectroscopy is a technique for the qualitative study of solid and liquid substances, of amorphous, crystalline, and gaseous types. In general, the analysis requires a monochromatic laser source, a system for beaming the laser radiation onto the sample and collecting the Raman signal, a system for separating the Raman signal from the elastically backscattered component, an interferometer, for the simultaneous acquisition of the entire spectrum, and a detector. The total spectrum, known as the Raman spectrum, is presented as a sequence of peaks and provides a digital fingerprint of the molecule in question, providing its identification by means of a comparison with reference database spectra (Fig. 4.6).

Raman spectroscopy has found numerous applications in the field of forensic science, because it has proven effective versatility in analyzing a wide range of materials, including also unknown and potentially dangerous substances, as it allows for analysis to be conducted through a glass or plastic container, and the great advantage of not destroying the sample at the analytical stage and not requiring preventive treatments that may alter it.

In the field of gemology, Raman spectroscopy is widely used to distinguish natural and synthetic gems, as well as detecting the treatment to which the precious

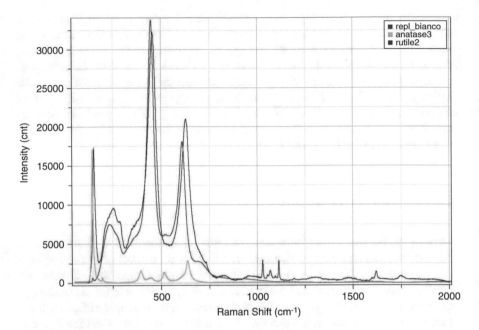

Fig. 4.6 Raman spectrum of a man-made fragment found in a soil sample, consisted in anatase and rutile. On the abscissa are represented the values of the Raman shift expressed in cm^{-1}, on the ordinates the values of the intensity of Raman signal

stones have been subjected in order to enhance their characteristics; for example, to discriminate whether natural diamonds have been treated in order to improve the color or transparency and hence inflate their commercial value. Furthermore, Raman spectroscopy is used to study and identify gems inclusions and any treatments applied to hide imperfections, which have a strong impact on their value.

In the field of mineralogy, it allows rapid identification of mineral phases with similar chemical compositions and crystal structures and to accurately identify the individual mineral phases in samples with heterogeneous mineral content.

4.10 FT-IR Spectroscopy

Fourier transform processing applied to Infra-Red absorption spectroscopy (FT-IR) is a technique that enables the study of the chemical bonds of organic and inorganic materials and then to identify them uniquely. The technique is based on the principle that when an infrared photon is absorbed by a molecule, this passes from a ground state to an excited vibrational state causing a transition between vibrational energy levels is a function of the actual absorption intensity. FT-IR analysis consists of irradiating the sample with a beam of infrared light in a wavelength range between 4000 and 400 cm^{-1}, in recording, after selection with appropriate optical systems, the absorption of individual frequencies, in the form of graph.

The spectroscopic study of absorbed frequencies in the infrared band can provide information about the functional groups present in the molecules that make up the sample and thus, indirectly, about the molecules themselves. In fact, the spectrum obtained by plotting the intensity of the absorption as a function of wavelength, although it is referred to the molecule in its entirety, is characterized by peaks associated with specific functional groups, forming part of its structure. Thanks to the reproducibility of these peaks, and especially the characteristic values of absorption, it is possible to reconstruct the structure of the molecule under examination. Information obtained by means of infrared spectroscopy is mainly qualitative, while the quantitative aspect is most widely used in the study of atmospheric pollutants.

The versatility of Fourier transform infrared absorption spectroscopy analytic techniques sustains diverse applications; in fact, depending on the chosen methodology, it is possible to analyze material that may be organic or inorganic, crystalline or amorphous and in any physical state (solid, liquid and gaseous).

FT-IR spectroscopy presents numerous advantages when applied in the forensic field, including the small sample quantities required, the ability to analyze the sample in its natural state, the speed of the analysis and acquisition of structural information, useful for the immediate recognition of broad classes of organic and inorganic compounds.

Case Report

The case illustrated below refers to the theft of a van, which took place in 2010 in Catania, during which the owner, in an attempt to thwart the robbers, was run over and killed by the van itself. The robbers, promptly tracked down and arrested by the police, claimed that the owner, when trying to stop them, had clung to the door of the car and then fallen sideways on the road. The investigators, however, suspected that the robbers had been determined to get rid of the van's owner had deliberately run him over, causing his death. A clear event narrative was crucial to specify the criminal charge: if the owner of the vehicle had fallen to the side on the road the robbers would have been charged with manslaughter, on the contrary, if they had run him over, the charge would have been murder.

The victim's clothing showed many traces of fine-grained dark material. The largest and most significant traces were found on the trousers; they were mapped and interpreted to formulate a hypothesis for the position of the victim at the time when that material had been transferred onto his garments. Although the removal of the particles from the clothing proved difficult given their dispersal among the fabric filaments, some traces, albeit with very small quantities, were acquired; however, the right-hand side of the trouser leather belt had a dark patch with plenty of fine trace material that was easily removed for further analysis (Fig. 4.7). Samples were taken from the van's mechanical parts which hypothetically could have come into contact with the victim had he been run over, in order to compare them with the tracks taken from the trouser belt.

The analysis showed that traces of material taken from the victim's belt and the samples from the mechanical parts of the van had similar color, particle size and morphology characteristics. In particular, all the samples showed various aggregates of very fine particles soaked in a dark oily substance.

Fig. 4.7 The dark fine
traces material on the
victim's trousers belt

The x-ray diffraction study indicated that the mineralogical composition of all samples was very similar. In addition to this, there was a particularly notable complete overlap of the crystalline phases present in the traces taken from the victim's belt, with those of samples taken from the mechanical parts of the van, consisting of quartz, calcite and dolomite and magnesium.

The dark colored aggregates on the belt and on the mechanical parts of the van were analyzed by FT-IR spectroscopy in order to determine the nature of the oily substance. The use of this technique showed that all the samples had lubricating oil and consisted of quartz and calcite, data, the latter, in line with those obtained with x-ray diffractometry.

In summary, the technical analysis allowed for it to be established that the samples taken from the victim's belt and from the mechanical parts of the van had very similar color, morphological, textural and mineralogical characteristics. Moreover, these samples had the same oily substance, identified as a lubricant for mechanical parts.

The strong comparability between the samples suggested that a contact had occurred between the mechanical parts of the van and the victim's clothing, sustaining the hypothesis of the run-over, during which the transfer of material had occurred. The distribution and geometry of the stains on the trousers also sustained the speculation that the victim was in lying position on the left side, coming into contact with the mechanical parts of the van with the right side.

4.11 Thermal Analysis

Thermal analysis allows for the measurement of the change of a physical property of a substance, for example the mass, as a function of temperature, while the same substance is subjected to a controlled heating cycle. The instrument is composed of four main parts: an extremely precise thermal weigh-scales, a furnace in which temperatures exceed 1000 °C, a gas purging system, which ensures a controlled atmosphere in the oven, and an interfaced computer which records and processes the analytical data.

Thermal analysis requires the use of a few milligrams of sample material, however, this is a destructive technique, that is to say that once a sample is exposed to high temperatures, its chemical and physical properties change; then the sample is no longer usable for further laboratory tests.

The thermo-analytical techniques have many applications and have been used in the forensic study of soils for several years. Their use in forensic geology can be an effective support for other analytical techniques such as x-ray diffractometry and chemical analysis.

Thermo-differential (DTA) and thermo-gravimetry (TG) analysis are the main thermal analytic techniques used in forensic geology. Additional thermal analysis, widely used in mineralogy, are emitted gases analysis (EGA), differential scanning calorimetry (DSC) and the thermo-dilatometry (TDA).

Thermo-differential analysis (DTA) measures the differences of temperature between a substance and a thermally inert reference sample during a controlled temperature cycle. Temperature differences depend on the thermal properties of the substance and the enthalpy differences caused by the chemical-physical transformations of the substance when it is subjected to temperature increase.

Thermo-differential analysis is particularly useful in the study of endothermic and exothermic phase transitions transformations. This analysis provides a spectrum in which the ordinate axis indicates the differences in temperature, while the abscissa axis depicts increasing temperature or time. Endothermic reactions take place when the instrument records a negative peak, while exothermic reactions occur with a positive peak. The geometry of the peak and the temperature range within which it develops allow for identification of the test substance by comparison with standard samples. Furthermore, the proportionality between the peak area and the amount of reactant material, allows quantitative indications to be obtained. Endothermic reactions are related to the loss of water, adsorbed or structural, the loss of OH from clay minerals, loss of CO_2 from carbonates, loss of SO_3 from sulphates, and from phase transitions and melting or evaporation processes. Exothermic reactions are due to the combustion of organic matter, the oxidation of compounds with divalent iron or manganese, to re-crystallizations of products and to structural transformations during cooling.

Thermogravimetric analysis (TG) is a technique that measures the changes in weight of a substance as a function of temperature and time. This analysis provides a thermo-gravimetric curve that shows the weight variation as a function of temperature and gives information on the stability of the sample and of the components that may be formed during the heating process.

In general, a thermogravimetric instrument also gives the derivative of the thermo gravimetric curve (DTG). This allows for the identification of the value of the temperature at the beginning of a reaction and the temperature at which the change of weight of the sample has the maximum rate of change.

   Thermal analysis applied to the study of soil samples can provide both qualitative and quantitative data on clay minerals, carbonates, sulphates and organic components. For example, the peaks of an exothermic reaction due to combustion of plant fragments, will change depending on the type of plant, their abundance in the ground and of their state of dehydration or decomposition. In addition, thermal analysis can provide information on the presence of fertilizers in the soil, humus and humic acids and their degree of maturity.

Chapter 5
Remote Sensing Applications in Forensic Investigations

Maurizio D'Orefice and Roberto Graciotti

Abstract Remote sensing applications in forensic investigations can be used as a forensic investigative tool, since it can provide data to the various judicial authorities, ordinary, criminal and civil, mostly engaged in the investigation of crimes against the environment and the territory. In this chapter the main features will be explained both theoretically and practically thanks to some case studies.

Keywords Remote sensing • Aerial photography • Landfill • Quarry • Forensic environment

The observation and interpretation of aerial photographs in stereoscopic vision is an investigative technique, known as photointerpretation, which has found a place in some Earth Science areas, particularly in the geological, geomorphological, and topographical fields.

The term geological photointerpretation or photo-geology has, in fact, come into common use and in several academic courses this analysis technique has been included as a course subject.

The technical photointerpreter observes aerial photos with special optical instruments, defined stereoscopes, that enable stereoscopic vision and the rapidly acquisition of information on the physical and geo-environmental context of a specific area.

Through stereoscopic observation it is possible to detect, in three dimensions, the main forms of erosion and accumulation that characterize a given area and, where possible, the structural setting of outcropping lithotypes.

The simultaneity of aerial photographs, also allows for fast response times in a territorial overview of the case, for example, of catastrophic events such as landslides, floods, volcanic eruptions, earthquakes and so on.

M. D'Orefice (✉) • R. Graciotti
ISPRA – Institute for Environmental Protection and Research, Rome, Italy
e-mail: maurizio.dorefice@isprambiente.it; roberto.graciotti@isprambiente.it

© Springer International Publishing AG 2017
R.M. Di Maggio, P.M. Barone (eds.), *Geoscientists at Crime Scenes*,
Soil Forensics, DOI 10.1007/978-3-319-58048-7_5

Below are specified the main features that make photointerpretation a technically very effective, and in some ways unique, analytical technique.

- According to the chosen flight altitude, stereoscopic viewing of aerial photos allows for a synoptic and comprehensive vision of areas that may also be very extensive and makes possible the identification of the main anthropogenic factors and their impact on the territory. We can mention, for example, quarries, installations for the disposal of waste of various kinds, important infrastructure such as bridges, dams, highways, the urban setting of expanding settlements, archaeological sites, etc.
- Photointerpretation can assist the detection, with sufficient accuracy and in a short time, of the physical characteristics of the examined area such as hills, mountain ranges, rivers, lakes, rocky peaks, plains, etc. and its related vegetation, reducing reconnaissance time on the ground and the associated costs.
- By photointerpretation, it is possible to observe the ground "at a distance", i.e. without direct operator interaction with the area under investigation. The use of this methodology is particularly useful when it is difficult, and sometimes impossible, to directly access in the areas under investigation for direct terrain inspections. Consider, for example, areas subject to active mining activity, particularly hazardous industrial sites, areas used for the disposal of toxic and harmful substances, inaccessible and elevated areas.

Accordingly, it is evident that photo-interpretation can be used as a forensic investigative tool, since it can provide data to the various judicial authorities, ordinary, criminal and civil, mostly engaged in the investigation of crimes against the environment and the territory.

5.1 Photointerpretation

Before exploring the main technical aspects and various operational phases regarding the development of an analysis of photointerpretation and describing also its applications in the forensic field, it is necessary to point out that this methodology for territorial investigation must never be confused with remote sensing and photogrammetry.

The latter are of particular technical and scientific disciplines, of which photointerpretation may be considered a derived application, having in common with them aerial photos as a basic tool.

Remote sensing acquires, by means of special sensors placed on satellites, aircraft and drones, geo-environmental data and information related to the earth's surface by measuring electromagnetic radiation emitted, transmitted or reflected by objects and elements on the ground in the area to be investigated.

Photogrammetry processes the aerial images through appropriate optical instruments and programs with high plano-altimetric precision (sometimes to centimeter resolutions), "returning" the shape, size and position of objects on the Earth's surface. It is used to make high-accuracy topographic and geothematic maps at various levels of detail.

These scientific disciplines, born in Europe in the mid-twentieth century especially for military purposes, have also been gradually emerging in other fields such as archeology, architecture, geology and geomorphology, forest sciences, and environmental engineering.

Photogrammetry, according to the picture-taking technique (aerial photos), is divided into two subcategories:

- Aerial photogrammetry: the photography employs cameras installed on aircraft (airplanes, helicopters, drones, etc.);
- Terrestrial photogrammetry: the photography employs ground cameras mounted on dedicated support gear. It is mainly used for architectural and archaeological purposes.

The required images are captured by means of cameras, these days of the digital type.

The processing and "rendering" of the images is performed through the use of complex digital tools and dedicated software. These computerised techniques fix the coordinates of all the points present in every single aerial photograph, enabling subsequent rendering in a geo-referenced scale for the generation of topographic maps.

The term photointerpretation denotes an investigative technique that provides qualitative and/or semi quantitative information on the physical structure of a given territory by means of the stereoscopic observation of remote sensed photographic images acquired by satellite, aircraft, drones, etc., or from ground-based equipment.

Geological and geomorphological studies make extensive use of aerial photos taken from heights of between a few hundred meters and 3000–4000 m in altitude.

Ultimately, photogrammetry and photo-interpretation are two methods of investigation using remote sensing photographic images as a basic tool. Photogrammetry processes the frames with analogue, analytical and digital instruments to achieve the "rendering" of topographic maps with very high plano-altimetric precision. On the other hand, photointerpretation analyzes photo images with more simple optical instruments, known as stereoscopes, to acquire non-numeric data on the physical characteristics of a specific area.

5.1.1 Aerial Photos

Aerial photography provides an image of the terrain that can be defined as a central projection of the object photographed on a plane.

According to the type of captured image, aerial photos can be distinguished as:

- *Panchromatic black and white*. With different shades of grey, they depict what in nature is diversely colored. These films have a low sensitivity to green light.
- *Infrared black and white*. These use shades of grey to represent objects capable of reflecting infrared light. Since water completely absorbs this type of light, in an aerial photo, waterways, wetlands and aquatic bodies appear black.
- *Color photos*. These reproduce objects with their real colors.
- *Infrared or false color photos*. They are mainly used in studies of the vegetation, when it is required to highlight some details of tree species, which appear in these photographs with specific and sharp colors.

The legal authorizations to conduct air flights for aerial surveys are issued by relevant departments for flight safety of individual nations. In Italy, the permits are issued by the General Staff of the Air Force.

5.1.2 Flight Techniques for Obtaining Photographic Strips

The aerial images are captured by a camera, usually digital, with its optical axis set perpendicular to the earth's surface. The camera is fitted to the fuselage of an aircraft flying at altitudes variable from hundreds to thousands of meters. The camera has particular characteristics regarding the lens focal length and aperture. It has a focus fixed at infinity, an automatic shutter trigger mechanism, special suspensions to limit the effects of vibration of the plane, a large negative format, which allows large print formats, typically 23×23 cm, and maximum precision optics.

During its previously planned straight-line route, the plane flies over the territory at constant speed and altitude. The shutter trigger sequence occurs automatically and in such a way that each frame includes about 60% of the territory captured in the previous frame (Fig. 5.1). The area common to the two images is defined as the longitudinal overlapping area (*overlap*).

What is obtained strip is a sequence of continuous superimposed frames, triggered at regular intervals along the same route. More strips with parallel routes must have a side overlap (*sidelap*) variable at least between 10 and 20%. This is necessary to ensure that all the territory to be investigated is covered by frames.

The frame represents an aerial photograph on which it is possible to identify the central perspective point through the use of special tracks (crosses or triangular notches), defined as reference marks, positioned at the four corners of the aerial photo and/or in the middle of each side of the same. By joining together the reference marks, it is possible to identify the main point of the frame (Fig. 5.2).

The flight line is the route along which two consecutive frames have been triggered and is obtained by joining the main points of each frame.

The air base is the distance that the aircraft covers between two successive frames and corresponds to the distance between P1 and P2 in Fig. 5.1.

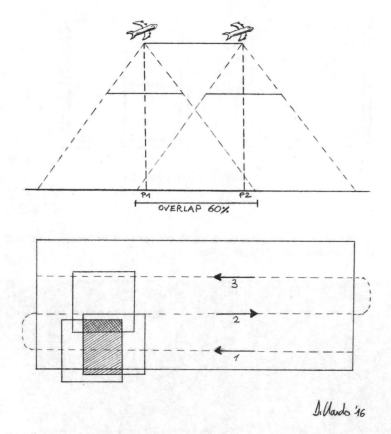

Fig. 5.1 The aerial photographic technique. Each frame of a given strip must overlap with the next one (area shown by the *oblique line*) to the extent of about 60% (*overlap*). Each strip, indicated by the numbers 1–3, must overlap the adjacent one (area highlighted by the double *oblique line*) to the extent of about 10 ÷ 20% (*sidelap*)

Each frame has imprinted on one side an image that depicts the status of the on-board instruments, defined as the data strip (Fig. 5.3). The elements of the data strip are:

- a circular level central bubble that controls the aircraft's horizontal trim;
- a clock that shows the time of the aerial photograph;
- the day, month and year of the flight;
- an altimeter that provides the flight altitude expressed in meters or feet;
- a plate with the serial number of the camera and the focal length of the lens;
- the sequence number of the frame.

It is precisely these characteristic elements on the side of each frame that make it a key document to be used in the forensic field. In particular, it is of great importance the recording date of the aerial photograph that certifies the time, day, month and year in which it is performed.

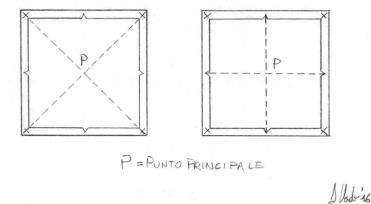

P = PUNTO PRINCIPALE

Fig. 5.2 By joining together the reference marks, the main point of the frame is identified (P). The line joining the two main points is the flight line

Fig. 5.3 Elements that constitute the data strip always present on the edge of a frame

Thanks for the information in the data strip, every frame is a snapshot of the "ground reality" and the human activities depicted on it reference a precise moment in time.

As will be specified later, the data strip is important when, for example, it is required to take into account the urban context in a specific area, to check for any unauthorized buildings constructed after the entry into force of any law or after the approval of a local zoning plan.

5.1.3   The Scale of the Frame

The focal length of the lens and the flight altitude are very important data for the determination of the scale of the frame. In fact, these allow to estimate with a reasonable accuracy, the size of the objects present on the frame and, more generally, the dimensions of the represented terrain. However, consider that:

- The frame of the scale is uniform for the areas around the main point of the frame and slightly different proceeding towards the edges of the same frame.
- The detected scale must always be understood as an average scale. It is, also, more accurate when depicting a flat landscape compared to rugged, undulating and sloping terrain.

There are two methods for determining the average scale of a frame:

- the relative altitude and focal distance method;
- comparison method relating aerial photos and the topographic base.

The first method uses the focal length (L_f) of the camera's lens and the flight altitude. These two elements are present on the data strip. To determine the relative altitude (A_{rel}), simply subtract the mean height above sea level (h.a.s.l.) of the terrain depicted in the frame from the flight altitude.

The denominator of the scale 1/X is the ratio:

$$X = A_{rel} / L_f$$

For example, if the aircraft flies at a relative altitude (A_{rel}) of about 2000 m and the focal length of the camera (L_f) is 152 mm, the average scale of the frame will be: X = 2000 m/0,152 m = 13,157; and therefore approximately 1:13,000.

The second method is based on determining the distance between two known points, which must be well identifiable both on the frame (D_{fot}), as well as on the topographic base map (D_{top}). These points, usually coinciding with man-made elements, must be at similar heights and at a distance of at least 5–10 cm, near the center of the frame.

The denominator of the scale 1/X is the ratio:

$$X = D_{top} / D_{fot}$$

For example, if the distance between two known points, measured directly on the frame (D_{fot}), is about 70 mm and the distance between the same points on the topographic base map (D_{top}) (for example a 1:25,000 scale map) is 20 mm:

X = (20 × 25.000)/70 = 7,142; the average scale of the frame will therefore be approximately 1:7,000.

5.1.4 Stereoscopes

On a frame which lies on a two-dimensional plane, it is possible to measure only two coordinates X and Y. If the represented area is shot from two different angles, under certain conditions a three-dimensional view can be reconstructed with the third dimension, Z.

To create an image that is stereoscopic and, therefore, three-dimensional, it is necessary that between two frames there is a longitudinal overlap of about 60%; in practice the same portion of territory must be common to two consecutive frames. In addition, the optical axis of the camera must lie on the same plane and the two frames must have the same scale. Some tolerances is possible on these values, but if the two scales differ by more than 15% it is impossible to have a stereoscopic image.

A pair of frames depicting the same object with the manner described above is a stereogram, also known as a *stereoscopic pair* or *stereoscopic model*.

The stereoscope is an optical instrument that allows stereoscopic images to be indirect viewed by means of a converging lens system, allowing for the simultaneous observation of two distinct frames providing a three-dimensional image. The stereoscope, in practice, directs the optical axes of the eyes to the corresponding points of the frame pair.

The first stereoscopes were made in 1832 by *Sir Charles Wheatstone*. In the market there are currently various types of stereoscopes, from the most simple and very cheap, to those used for teaching, which allow for simultaneous vision by a pair of operators, to those connected to a PC using special software.

Here we will limit ourselves to describing the more readily available simple types on the market:

- lense-based stereoscopes;
- lense and mirrors based stereoscopes;
- lense, prism and mirrors single or double based desk-top stereoscopes.

The first type comprises only of two plano-convex lenses, with the upper face flat, mounted on a suitable metallic or plastic support. These are adopted for campaign photo-interpretation activities due to their small size and ease of transport (generally pocket-sized). Their main drawback is that stereoscopic vision is not allowed for the entire overlap but is limited to small portions of terrain. In view of the size of the aerial photos (23 × 23 cm), and taking into consideration the interpupillary distance of the tool (on average around 65 mm), it is less than the distance between two homologous points on the stereoscopic pair (about 20 cm), during examination of the model it is necessary to overlap and fold a single frame.

Stereoscopes based on lenses, prisms and mirrors (Fig. 5.4) can be desktop, single or double (for teaching), or portable. They allow detailed analyzes and accurate observations of the details as they are equipped with interchangeable lenses in two magnifications: one between 0.7× and 1×, the other between 3× and 8×. The pupillary distance can be adjusted between 50 and 75 mm. The field of stereoscopic observation is wide and allows the examination of the entire area of overlap of the two frames. The instrument is also high enough to allow the photo-interpreter to easily draw on the frames the various functional symbols useful to its analysis.

Fig. 5.4 Some types of portable and desk-top stereoscopes

5.1.5 The Technique of Photointerpretation

To obtain satisfactory results, the photointerpretative analysis must be conducted in accordance with specific operational steps and simple measures.

- Collect all available flights for the area under investigation and the various topographical and geothematic maps at different scales of detail. Photointerpretative analysis should examine aerial photographs starting from the earliest flights.
- Ensure good stereoscopic vision in the selected frame pairs (model pairs). The two aerial photos are arranged under the stereoscope in the same orientation that they had at the time of shooting. This is achieved by determining the main point of each frame by means of the intersection of the reference marks and then uniting the points obtained in a direction parallel to the line joining the optical centers of the two lenses. The distance between corresponding points on the frames is a constant of the stereoscope and generally is about 24–25 cm.
- During the stereoscope observations, it is essential to remember that the image is exaggerated in the third Z dimension compared to ground reality. It is therefore advisable to compare aerial photos with the basic topographic map to estimate the real value of the altitudes and then correct the stereoscopic exaggeration.
- After an initial careful, skilled, and synoptic observation of the terrain, it is useful to outline areas that are homogeneous in shape and size and match the items with similar characteristics.
- In general, the anthropogenic forms have regular features, sometimes with curvatures, while the elements and the forms that characterize the natural terrestrial

physical environment have more irregular trend, with variable and articulated contour boundaries.

- For good results, it is crucial that the photointerpreter performs the checks directly on the ground to check the exact correspondence with the measured data.

The results obtainable by photointerpretation, especially if addressing forensic field investigations for the acquisition of evidence in judicial investigations, are closely related to the technical skill of the photointerpreter. This skill is acquired only through long experience of observation accompanied by frequent campaign inspections.

Photointerpretation is a valid technical tool to be employed before, during and at the end of the survey operations.

It is to be noted that this methodology is a tool in support of the other investigations and will never fully replace direct land survey carried out by instrumental and laboratory analyzes.

Photointerpretation can be conducted, as a preliminary measure, through the summary photo-reading of a single frame, without the stereoscopic vision and in a thorough manner through the stereoscopic vision of a pair of frames.

The photo-reading of a single frame, allows you to get information about:

- the general degree of the human activity, particularly with regard to the state of urban expansion;
- the presence of quarries and waste disposal sites;
- the existence of important infrastructure such as dams, bridges, airports, etc.;
- the layout of the primary communication routes, with particular reference to the main roads, byroads and paths;
- the type of vegetation cover, to distinguish areas with woodland, fields and fallow;
- the overall layout of the surface hydrographic network.

The stereoscopic view of a pair of frames allows us to acquire information about:

- the hydrographic structure and topography of the land and the steepness of the slopes;
- the configuration of morphological elements that characterize the landscape;
- zones the difficult to access at high elevations or with steep slopes.

Photointerpretation will be full and detailed If executed on images obtained from flights taken at different times and at different scales. In these cases, we use the relevant photointerpretation terms *multitemporal* and *multiscalar* respectively.

5.1.6 Multitemporal Photointerpretation

If an area has been subjected to a number of flights in different periods, the comparison of the various frames will highlight any changes in a given area over a specific time interval.

Fig. 5.5 Example of multitemporal photointerpretation. The two frames depict Golfo di Campo with the village of Marina di Campo (Island of Elba – Italy). The left frame was taken in January 1944, the right one in March 1998. From the comparison of the two pictures, even at different scales, it is easily to assess the building expansion that has forced profound changes on the coastline

This is possible because the data strip present in every single frame records the year, month, day and time when the recording was made. Multitemporal observation is also very useful for estimating the evolution of the natural landscape and the anthropogenic context of the area under investigation (Fig. 5.5). Of course, the study of the evolution of a territory will be all the more detailed the more numerous are the available flights and the longer the time period covered by them. Using the oldest historical flight material (e.g. in Italy the 1943 RAF flights or the GAI flights – see Appendix), and having data from flights made in recent years, it is possible to estimate territorial evolution over the past 60–70 years. As we will see later, this is especially useful if investigations are required on illegal construction and control of mining activities.

5.1.7 Multiscalar Photo-Interpretation

If a given area has been subjected to numerous fly-overs, performed at different altitudes, then the higher altitude images will enable panoramic and general observations, while the lower altitude flights will provide more accurate and detailed observations. General views from high altitudes will detect natural and man-made macroforms and dominated by significantly linear shapes. Low altitude images, on the other hand, provide details on shapes of smaller forms (Fig. 5.6).

Fig. 5.6 The two aerial photographs represent a part of the Foro Italico – Rome. The frame at the top (**a**) was taken at a relative flight altitude of about 2000 m; the one at the bottom (**b**) at the relative flight altitude of about 500 m. There is an evident greater amount of detail associated with the lower photo (Source: Cassinis and Solaini, 1946)

Note that some stereoscopes are equipped with interchangeable eyepieces objectives that allow for a "zoom effect" that can be significant. Also in these cases, details in the frames will be observable with strong magnification but with results still inferior to those achievable with low-altitude flights.

5.2   Aerial Photo Parameters

During photo-interpretative analysis, some parameters may be considered, typical for a same photo, which provide, in an indirect way, information about the characteristics of the examined terrain. These parameters generally vary every time there is a change in the perspective angle with which the camera photographs the land surface. Such differences may also be detected between adjacent frames of the same strip. Below is a brief description of the main parameters to be examined in a photointerpretative investigation.

5.2.1   Grey Tone

By analyzing the shades of grey in a black-and-white photograph, useful information can be obtained from the diversity of the shades themselves. The grey tone can be influenced by several factors such as morphology, lighting and shadows.

In general, aerial photographs dominated by the same shade of grey can be considered homogeneous (Fig. 5.7).

Land is directly influenced by the amount of water content; the greater the degree of humidity, the darker is the grey content, up to the full black denoting surface water. Dry soils are characterized by very light shades.

Soils high in organic matter are darker than impoverished ones. Sialic beach sands appear white when dry, however, when wet they show up as dark-grey.

The lithological nature of a rock is closely linked to its mineralogical component which influences, in turn, the tones of grey in a black and white panchromatic aerial photo. For example, intrusive igneous rocks with high content of sialic minerals (quartz, feldspar, muscovite, plagioclase, etc.) in images frames usually show light grey tones. Whereas effusive rocks rich in femic minerals (pyroxene, biotite, amphibole, olivine, etc.) project darker grey tones. The limestone rock types have varying shades of grey (there are normally light grey), sandstones generally appear as light grey and clays appear as dark grey.

Fig. 5.7 Detail of an aerial photo that depicts an extensive flood plain intensely affected by cultivated fields and crossed by an irrigation canal that runs through the frame from north to south. The
individual fields (crops) have very regular geometric shapes that indicate their evident man-made
origin and several grey tones from white up to black

5.2.2 Texture

Texture in a frame is portrayed by micro-changes in the distribution of the different
shades of grey in a specific area.

The tone can present considerable variations between continuous areas, sometimes, however, its distribution is more uniform across the entire frame.

A complex and articulated terrain morphology will give rise in the aerial photo
to a heterogeneous and uneven texture because the darker tones, due to the shadows
caused by scarps and slope discontinuities, will have a random distribution. An area
characterized by a flat surface topography, with a constant slope, in the photo will
present lighter tones alternating with darker ones and, consequently, the texture will
be uniform and regular (Fig. 5.8).

Fig. 5.8 Excerpt from frame nr. 1741 in the GAI flight archive, taken 2 September 1954 over the Laga mountains (municipality of Acquasanta Terme – AP – Italy). The slopes are evident with their characteristic light and dark bands caused by alternating sandstone and pelitic layers dating back to the Laga formations

5.2.3   Shadows

Depending on the season and time of acquisition, the frames may show areas with greater or lesser degrees of shadow. Shadows can be originated by rugged and uneven terrain (deep narrow valleys, sub-vertical cliffs scarps and rocky peaks), by forest cover or urbanized areas.

Shadows, when they darken large areas, can be a hindrance to photointerpretation, but sometimes they are able to provide useful information and important clues to the photointerpreter in the evaluation of some elements. Shadows may in fact be useful for reconstructing the vertical development of some elements (in the third Z dimension) such as buildings, trees or orographic reliefs.

5.2.4   Structure or Pattern

The structure of a frame is concerned with the distribution of different objects and elements in the area. This parameter is closely related to and affected by the geological, geomorphological and hydrographic context of the area in question.

In an aerial photo, surface hydrography (drainage) is easily recognizable because the channelled water shows very dark shades of grey.

These channelled water, as a function of their density, the type of confluence and distribution, assume particular configurations known as *drainage patterns*, or the hydrographic grid.

The different types of pattern depend on various parameters, among which the most significant are the lithological composition of the outcropping rocks, different erodibility and permeability of the rock types, the structural and tectonic geomorphology of the area.

Fine-grained and highly impermeable soils and have particularly articulated and developed drainage patterns. Conversely, permeable karstic limestone characterized by fissures and fractures and inconsistent porous highly permeable terrains have relatively undeveloped drainage patterns.

Easily eroded rocks have a more developed and articulated network than lithoid types that are compact and less erodible.

Tectonic rock formations (with faults, fractures, etc.) are generally characterized by relatively straight river courses, aligned according to one or more preferential directions.

5.3 Forensic Photointerpretation

Following what was illustrated previously, it is evident that the use of photointerpretation as an investigative technique is particularly suitable when working in the forensic field. This is especially true in the case of findings and insights arising from the work of Authority Technical Experts, private Legal Technical Consultants, or that of Investigative Technical Experts engaged in counteracting crimes against the environment and the territory.

It must be emphasized that the possibility of observing a territory remotely without direct interaction with the sites of investigative interest is a prerequisite, especially when accessing the areas under examination is difficult, and sometimes impossible. This happens especially when the research involves quarry areas, industrial facilities subject to special safety restrictions, illegal waste disposal sites.

The opportunity to have, via aerial photos, a synoptic view of even very large areas, allows the appointed technical expert to reduce the investigation time and concentrating his work where it can yield the best results. This occurs, for example, when investigating unauthorized construction activity within national parks and/or protected areas or when considering the interactions and environmental impacts affecting infrastructure and the physical environment.

Finally, it is important to note that aerial photos are effective documents to be used in the course of criminal investigations since, using the information derived from the data strip, they represent the "ground reality" at a precise moment in time (hour, day, month and year).

There is not enough scope in this text to cover in detail the many applications of photointerpretation in the forensic field. Thus, by way of example, the following are some cases studies referred to the Italian territory.

5.3.1   Mining and Quarrying

In Italy, the mining sector is regulated directly by the State, while quarry extraction is delegated to the regional authorities that are in charge of individual permits and the quantity and type of materials extracted.

The presence of quarries on a given territory triggers a series of high-impact geoenvironmental criticalities.

In these cases, landscape undergoes profound changes that are gradually more and more pronounced as can be seen in Fig. 5.9, which compares an excerpt of one frame from a 1954 GAI flight and a 2015 Google Earth image of an area in south-central Italy. In particular, in the 1954 frame note how the limestone relief, in contact with the flood plain, is partly affected by some quarries (recognisable in the aerial photos as small white spots with rounded outlines) located in the south-western and north-eastern areas. In the 2015 image, the quarries have spread all over the sector, the largest of them is located in the center and has the typical stepped slopes of a quarry. It is also interesting to point out the intense urban development.

The transformations induced by mining the landscape are intense to the point that entire hilly areas have been flattened.

Hilly areas subjected to transformation over time reach an almost flat conformation. River bed extraction sites alter the longitudinal profile of river courses. Open quarry frontages on steep slopes deeply modify the morphology of entire rocky outcrops.

Mining activity also gives rise to a series of problems related to the pollution of groundwater, in particular, unregulated disposal of quarry sludge, intense heavy vehicle traffic on the road network surrounding the area, vast amounts of dust and strong noise.

These profound changes to the environment can be analyzed and monitored by the technique of multitemporal photointerpretation. For example, we can begin by examining the frames of the oldest flights to arrive at the most recent ones, it is possible to observe the changes suffered by the territory as a result of mining and verify if these changes that took place over several years (sometimes even several decades) have been carried out in compliance with prevailing regulations.

During their active life, and often for long subsequent periods, the changes leave deep "wounds" in the territory. Proper management of mining activities envisages remediation efforts and environmental restoration at the end of the productive period.

Photointerpretation can provide a useful aid in monitoring that these operations are undertaken properly, especially when these works are undertaken in places with access difficulties, such as, for example, shaping and landscaping the step terraces of the former quarry fronts.

Given the strong impact a quarry has in the territory and its great visibility in aerial photographs, photointerpretation is a very effective technique in investigations aimed at discovering illegal and uncontrolled mining activity. Sometimes there is evidence of open quarries in areas subject to restrictions or protected areas, or pit quarries made in flood plains or directly in river beds.

Fig. 5.9 Quarry areas in central and southern Italy: (**a**) extract from one frame relative to the 1954 GAI flight (**b**) the same area in a 2015 picture taken from Google Earth

Often the photointerpreter is invoked when criminal investigations are directed at the verification of planimetric calculation of the quarry extension and the height of its fronts, during and at the end of mining, to verify the correspondence of the extracted volumes to the provisions of the initial mining concession.

5.3.2   Unauthorized Building Construction

Photointerpretation is particularly appropriate when the appointed technical expert must carry out investigations regarding the illegal construction activity. This phenomenon, unfortunately very widespread, has a very negative impact on the natural landscape and the environment. Just think of the infrastructure and housing built indiscriminately and unlawfully, in protected coastal and river areas, in areas of high historical and landscape value (archaeology parks, nature reserves, protected areas, etc.) and in areas of high seismic, volcanic and hydrogeological hazard.

In this case, the photointerpreter, through the synoptic observation of a very large area, can identify and demarcate the protected and/or restricted areas on the frames and identify the structures and homes built illegally within them.

As an example, we can cite what happened in Italy after the entry into force of Law no. 47/85, which represents an important regulatory tool of reference in the field of unauthorized building. In fact, this law has allowed, for the first time and in a structured manner, the regularization of the legal positions of the owners who built illegally.

Photointerpretative analysis allows for the acquisition of the date (year, month and day) and time of the aerial photograph imprinted on the data strip and, therefore, to see if the flight has been carried out prior to or following the date of entry into force of applicable zoning laws. As a result, multitemporal observation allows, therefore, to assess the existence or otherwise of irregular construction activity.

This method is suitable, above all, for environmental monitoring activity referred to illegal construction within the territorial limits of areas that are protected and/or subjected to hydro-geological and environmental restrictions of various kinds. In fact, by analyzing aerial photographs taken before or after the entry into force of the different intended use regulations for the individual areas, it is possible, also in this case, to verify what has been implemented unlawfully.

5.3.3   Illegal Landfills

The Italian legislation on waste disposal describes the types of landfills, the types of waste treated according to their danger and also the technical standards for the proper management and implementation of disposal facilities.

A landfill is defined as "controlled" when the design, construction, disposal methods, and system monitoring are properly carried out in compliance with current regulations.

In these cases, the soil and subsoil will be protected from pollution risks associated with accidental releases of leachate by means of a natural geological barrier and a suitable artificial waterproofing of the bottom and the walls of the landfill.

In Italy, until recently, the disposal of waste in landfills has been the main method of waste processing, while at present, the tendency is to reduce as much as possible the amount of waste sent to landfill and we are moving towards waste sorting techniques, recycling, composting and incineration.

Unfortunately, the disposal of waste is not always performed in a "controlled" manner. Illegally created landfills are numerous, given the enormous volume of business that is associated with trafficking waste, especially when it comes to toxic and harmful material. This phenomenon, controlled by actual criminal organizations, is a social evil, which creates significant damage to the environment and threatens the health of the population.

The search for illegal waste disposal is a significant activity and is a subject of commitment on the part of Justice departments and Law Enforcement Agencies, with great deployment of personnel and equipment.

Geophysical technical tools and new research methods available today allow us to obtain good results, although limited and circumscribed to the area studied.

A significant contribution to these investigations can be provided by photointerpretation performed by experienced technicians.

As we saw earlier, an aerial photo, according to the flight altitude, can cover very large areas. A pair of frames observed with a stereoscope allows, in fact, a general three-dimensional view of an immediate and very wide area and, moreover, without direct interaction with the ground; requirement of primary importance in these cases.

The photointerpreter, during the observation of the stereoscopic model, will detect any abnormalities in the morphology of the terrain, these forms may be of anthropogenic nature and therefore generated by illegal waste deposits. When the mass of waste extends vertically, it is possible to detect this terrain irregularity, especially in flat areas. Sometimes, in fact, "stacked or banked" waste piles rise to several tens of meters above ground level and create small hills with peculiar shapes that can hardly be attributable to natural geomorphological processes. It is these positive morphological abnormalities that the photointerpreter can grasp and attributed to the presence of a possible illegal landifill.

When, however, the illegal disposal site is of the pit type, the photointerpreter can detect abnormal depressions with respect to the surrounding terrain. The negative morphological abnormality is documented by the very regular depression outline, usually with a pronounced perimeter (often rectangular). These pits may correspond to lakes originated from rain waters and/or, sadly, even relate to the emergence of the groundwater during digging operations.

When data from multiple fly-overs are available, covering a well-defined and sufficiently long time intervals, multitemporal photointerpretation can be used to assess the morphological evolution of these man-made forms, whether they are hills or depressions.

When illegal disposal activities cease, the accumulated waste, also for several meters above the ground level, may be covered by spontaneous vegetation or, as often happens, by artificial plantations made with the purpose of concealing the underlying waste mass. In these cases, the photointerpreter can detect tree species that are completely unrelated to the surrounding arboreal and vegetative context.

The same phenomenon can be observed with pit landfills, when abandoned, they may be levelled to the original ground level using filler soils that can be easily recognized by the photointerpreter.

The data and information acquired, once made available to the judicial authorities and the police, can be used to guide further campaigns for the investigation, detection and sampling in specific areas and well-defined locations.

Appendix: A Brief History of Aerial Photography in Italy

The first aerial photographs in Italy were taken from a balloon in 1897, while the first pictures from aeroplane as part of a military reconnaissance were taken in February 1912, during the Libyan war, from the plane of Captain C. M. Piazza. It was, however, during the First World War and the years immediately following that the airplane began to be used for aerial photography.

The first aerial photos of Rome to be taken from an aeroplane are the flight of Umberto Nistri (1919), who later (in the 1920s and 1930s) founded the first aerial survey company (Sara Nistri) still in business today (Fig. 5.10).

Fig. 5.10 Photogram from the Sara Nistri flight of 1934. In the central part of the image can be seen the Vatican City with the Basilica of Saint Peter (Authorization for Use for editorial purposes by the Central Institute for Cataloguing and Documentation – MiBACT and the British School at Rome – further reproduction and/or duplication by any means is forbidden)

5S8234 3PG 6ᵗʰ Jᴀɴ 1944/1030/F24/22 500

Fig. 5.11 Photogram nr. 5S8234 belonging to the "the RAF fund" preserved at the National Aerofototeca ICCD. The photo, taken January 6, 1944, highlights the historic center of Rome. In particular, in the lower right corner you can see the Colosseum and the archaeological area of the Palatine, the Roman Forum, the Celio and the Colle Oppio (use Permission for editorial purposes by the Central Institute for Cataloguing and Documentation – MiBACT and the British School at Rome – further reproduction and/or duplication by any means is forbidden)

The Military Geographical Institute (IGM), starting in 1925, executed a series of aero-photogrammetric surveys on the national territory, in order to update the topographic plates in the 1:25,000 scale. The photographs taken since 1945 are available on the interactive catalogue on the IGM[1] website.

During the Second World War, the belligerent air forces flew over the Italian territory, for purposes of reconnaissance and strategic planning. While photos of the Italian Royal Air Force and the Luftwaffe (characterized by the large frames in 30 × 30 cm formats) are still little known, those taken by the Anglo-Americans between

---

[1] www.igmi.org/prodotti

1943 and 1945, however, are well-known and renowned. Apparently, millions of frames were taken, in two sizes (18 × 24 and 24 × 24 cm). The group of photos remained in Italy (more than 800,000 aerial photos), collected in two large collections traditionally known as "the RAF fund" (Acronym for the British *Royal Air Force*) (Fig. 5.11) and "USAAF fund" (acronym for *United States Army Air Force*), it is now preserved in the National Aerofototeca (AFN) at the Central Catalogue Institute and Documentation (ICCD – Ministry of Cultural Heritage), based in Rome, Italy. At the same Institute are also stored frames from supplementary flights that cover, at different scale and discontinuously, the national territory during a period of time, ranging from the late 1800s to the present, and survey flights of various eras, scales and areas. These flights were carried out over several years, by several private companies operating in the field of aerial photography and aerial photogrammetry.

Later, under an American contract, a consortium of private companies was formed, called the Italian Air Group (GAI), that between 1954 and 1955, made the first systematic stereoscopic aerial coverage of Italy (the GAI flight, also called the basic flight).

At the national level, low altitude flight images are available for the production of topographic maps at medium-large scale. They are available in the official websites of the individual regions, where, among other things, is available information on the conditions for the acquisition of photogrammetric and mapping material.

Chapter 6
Forensic Geomorphology

Maurizio D'Orefice and Roberto Graciotti

Abstract Geomorphology establishes a complete picture of the Earth's landform characteristics of a given territory from the moment of its formation until the present time, and predicting from this the future evolution of space-time. In the forensic perspective, geomorphology, and specifically geomorphological surveying, often represents a basic cognitive approach to the framing of issues related to the interaction between both anthropogenic structures and grounds, both incorrect and/or illegal activities implemented against the environment.

Keywords Forensic geomorphology • Forensic environment • Landforms • Geomorphological survey • Landslide

6.1 Overview

Geomorphology, including the methodologies, and surveying techniques used in it, may be an important investigative tool in the forensic field, like other disciplines of earth sciences such as mineralogy, geochemistry, geology, geotechnics, hydrogeology, geophysics, etc.

Geomorphology can be defined as the science that studies the nature, history, genesis, distribution and mutual relations of the earth's landforms and deposits related to them. The main objective of this discipline is therefore to establish a complete picture of the geomorphological characteristics of a given territory from the moment of its formation until the present time, and predicting from this the future evolution of space-time.

In this perspective, geomorphology, and specifically geomorphological surveying, often represents a basic cognitive approach to the framing of issues related to the interaction between both anthropogenic structures and grounds, both incorrect and/or illegal activities implemented against the environment.

M. D'Orefice (✉) • R. Graciotti
ISPRA – Institute for Environmental Protection and Research, Rome, Italy
e-mail: maurizio.dorefice@isprambiente.it; roberto.graciotti@isprambiente.it

© Springer International Publishing AG 2017
R.M. Di Maggio, P.M. Barone (eds.), *Geoscientists at Crime Scenes*,
Soil Forensics, DOI 10.1007/978-3-319-58048-7_6

The geomorphologist, through a targeted geomorphological survey, can provide a useful interpretation of the physical earth and thus contribute to the research of the causes and responsibilities related to certain operating environments, which in principle can be traced to:

1. Performance deficits after anthropic works are carried out, produced as a result, also when combined with natural morphogenetic processes not sufficiently addressed during the preliminary surveying activities before the project phases. One typical example may be the differential settlement of the foundations of a building, or the pylons of a bridge resulting from improper drainage of runoff water on land with poor permeability, or injuries that may occur in the lining of tunnels, caused by gravitational processes of slopes that are not assessed comprehensively or that are underestimated in project phases.
2. Causal link between the damage caused to the structural works of various types and action to man-made boundary due to incompetence, negligence sometimes even intentional. Such as roads, buildings, tunnels, bridges, dams, etc., designed and built in areas unsuitable from a geomorphological perspective, or subject to restrictions.
3. Legal disputes between the contracting authority and the company performing the work about geological and geomorphological issues have, at times, not been evaluated fairly during the preliminary investigation, the preliminary planning stages. These issues are very often due to delays in carrying out works with consequent increases in the costs of the work.
4. Crimes against the environment and the territory caused, for example, the construction of quarries in the river beds, for the withdrawal of large amounts of gravel and sand. These illegal mining activities can alter the longitudinal equilibrium profile of the entire stream, resulting in engraving phenomena and deepening its bed, which can cause stability problems for artifacts in the river bed, or nearby (for example, of bridge piers, artificial embankments, channel outlets, buildings, etc.). In addition, the extraction of material along the river channel significantly decreases the sediment transport at sea, resulting in an increase in coastal marine erosion.

It is not always possible to determine exactly the limits between the first two operational contexts (performance deficits of a man-made structure and causal link between damage and anthropic action). Their limits may in fact be nuanced and sometimes related, as in the case of failure of an artificial river embankment underpowered and the subsequent flooding of the buildings located on the back of the work in the flood plain.

Research into the causes and responsibilities related to the above contexts necessarily involves estimates of the technical assessments from the geomorphologist when he is appointed as an advisor to judicial authorities or one of the parties to the proceedings of a lawsuit, or by the judicial police.

Such inspections and surveys, which are aimed at preparing the technical report to be delivered to those who have been appointed, must be carried out according to their own methods and techniques of geomorphological surveying. A correct approach for the direct acquisition of field data often provides a decisive contribution to the operating framework and settlement of litigation that has been filed.

It is useful to remember that the field survey criteria must be part of the cultural and technical norms and expectations of the professional geologist, especially if he has to draw up the preliminary geological report for the design of a given work (building construction, road embankments, tunnels, arrangement of unstable natural slopes, works of large environmental impact, landfills, drainage, abstraction, consolidation of land, etc.). Even during forensic investigations, the geologist in charge must apply the surveying techniques with care and professionalism, in order to provide all relevant information to the judicial proceeding and in order to avoid direct responsibility for any damage to property and persons.

To better understand the basic concepts of geomorphological surveying technique, here the general geomorphology principles are briefly recalled.

6.2 Principles and Concepts of Geomorphology

6.2.1 Agents and Morphogenetic Processes

Geomorphology, as previously mentioned, is the science that studies and interprets the origin, history, distribution and mutual relations of the forms that make up the earth's relief, and deposits related to them, in relation to agents and processes that have generated and modified them.

The phenomena that are defined as *morphogenetic agents*, are those that give origin to relief forms and control their evolution in space and time, through the activation of specific mechanisms, namely, *morphogenetic processes*.

Landforms are normally produced by a combination of different processes; however, in many cases one of these, called the main morphogenetic process, can be considered a main factor.

The morphogenetic agents can be divided into two main categories: *endogenous agents* and *exogenous agents*. The former includes volcanism and tectonic activity, whose morphogenetic activity occurs by means of *endogenous processes* such as the emplacement of lava flows, or the surface faulting. The seconds are often represented by fluids (air, water, ice), responsible for the activation of *exogenous processes* (e.g. fluvial process, glacial, karst, wind, marine, etc.).

A major factor in the activation of virtually all exogenous processes is represented by the force of gravity which, in the case of landslides, assumes the role of main morphogenetic agent.

6.2.2 Landforms

In geomorphology, shape or form can be defined as a spatial entity constituted by a surface, characterized by its own configuration, i.e. the three-dimensional appearance of a surface (for example, flat, wavy, etc.), which represents the detection of an erosive episode (form from erosion) or the final morphological expression of a sedimentary episode (form from accumulation).

As regards accumulation forms, in geomorphology it is fundamental to study not only the surface (three-dimensional entity, but devoid of "thickness") which delimits a given deposit, but also the deposit itself (volumetric entities).

In general, the shapes of erosion have a negative aspect, that is, are characterized by a concave profile, while those of accumulation have a positive configuration, i.e. a convex profile. It should, however, be stressed that this rule is not always valid; there are, in fact, convex profile erosion forms (such as a sheepskin rock (Fig. 6.1), a *biancana* (Fig. 6.2), a rounded ridge, etc.) and although rarer, accumulation forms with a concave profile (for example a depressed area in the flood plain, a scree slope at the foot of the slope of a "U-shaped Valley", an accumulation *glacis*, etc.).

Of course, the size of the forms can be very different. In fact, they can range from those of enormous continents or ocean basins, which extend for tens of millions of square kilometers, to those represented by tiny grooves, holes, scratches, ripples, etc., that with areal growth of square centimeters or even millimeters, affect both the rocks and the sediments.

Landforms are commonly classified according to the following criteria:

Fig. 6.1 Typical example of "sheepskin rock" formed by the erosive action (abrasion) of a glacier (the Lagorai Chain, Val di Fiemme, Province of Trento, Italy)

Fig. 6.2 Typical landscape of *biancane* (badlands). It is convex shaped and white, created by the erosion of surface runoff on clay soils devoid of vegetation (Province of Siena, Tuscany, Italy)

- *Morphographic criterion*. Qualitative description of landforms, based on the use of specific terms (valley, fan, slope, cave, etc.).
- *Morphogenetic criterion*. Definition of the origin and evolution of landforms in relation to endogenous or exogenous process that generated them (river valley, alluvial fan, tectonics slope, sea cave, etc.).
- *Morphometric criteria*. Quantitative description of the dimensions of the relief forms (edge of river terrace of more than 10 m in height, with alluvial fan slope less than 2 %, etc.) and analysis of the spatial properties of some morphological elements through the use of numerical parameters (for example, quantitative geomorphic analysis of river basins and drainage networks).
- *Morphodynamic criterion*. Definition of the state of activity of the processes responsible for the genesis of landforms. In common usage the activity status is reported, albeit incorrectly, also for the individual forms, that can be distinguished in two broad categories: *active* and *inactive*. Between active and inactive forms *dormant* forms are found. Given the importance the forensic field gives to the matter of the definition of the concept of a form's activity, with all the associated problems, these are more widely discussed in subsection 6.4.1.
- *Morphocronological criterion*. The definition of the relative age of the relief forms, based on geomorphological and stratigraphic correlation factors (related to surface deposits), or where possible, their geochronology age (in years dating from the present). The latter is achieved using specific methods and dating techniques, applied to the deposits associated with the landforms, or in special cases, directly to the surface of the landforms themselves.

6.3 Geomorphological Surveying

The geomorphological surveying of land consists in the systematic observation, analysis and interpretation of erosion and accumulation forms (including surface deposits associated with these), that characterize a given territory. This activity is also aimed at their cartographic representation through the use of appropriate symbols and line graphs reported on a topographic base at an appropriate scale.

The geomorphological survey can be performed, by and large, according to some main phases (D'Orefice and Graciotti 2015), which provide:

- A search of the literature and archives, as well as cartographic research, to collect all available data about the area to be investigated and the surrounding areas;
- the analysis of remotely sensed images (aerial photographs, satellite images, land resumed or drone) (cf. Chap. 5);
- direct field surveying;
- geognostic surveying, sampling, laboratory analysis, instrumental technical monitoring;
- processing, interpretation, and the return of all the acquired data;
- the drafting of the final technical report to be transmitted to the judicial authorities, lawyers and law enforcement agencies engaged in criminal investigations.

Among the technical steps to be performed, that of geomorphological field surveying is certainly the most challenging. In fact, the surveyor, mainly operating on the ground, must possess a certain attitude and spirit of adaptation in order to interact with the surrounding physical environment.

The field survey can be conducted via a sequence of steps which may include: preliminary surveys, direct and systematic surveying of land, synthesis and processing of field data.

6.3.1 Preliminary Inspections

In the absence of a previous general knowledge on geological and geomorphological features of the area to investigate, it is better to gather this information through expeditious and targeted inspections.

For this purpose, an analysis of aerial photographs and possibly of orthophoto maps concerning the area to be examined must be carried out. Photointerpretation will be the more complete the greater the availability of aerial photos of different ages and stairs. This activity will be carried out before, during and after their completion of land surveys.

Even the examination of topographic maps at different scales of detail, will provide useful information on the morphological context of the area to be investigated.

To obtain an overall view of geomorphological phenomena, it is often essential to extend the first observations of land also outside of the area to be surveyed.

6.3.2 *Direct and Systematic Surveying of Land*

As regards surveying methods (basically designed for the recognition of the main erosion and accumulation landforms and their surface deposits), it should be noted that, according to the hydrographic and orographic structure, the area to be investigated may fall within one or more main catchment basins. These in turn can be divided into an increasingly lower hierarchy of sub-basins.

Within each basin, the operational phase of land must be conducted in a systematic manner over the entire area to be investigated, along all roads, mule tracks and paths. At this stage, it is appropriate to make the observations not only along predetermined routes, but also outside of them, so as to fully cover the territory or, in case of inaccessible areas, try to investigate the greatest surface area possible.

As for the equipment necessary for landscape investigations, it is good to distinguish between essential equipment to be kept to hand during field operations (Table 6.1), and that which is only to be used for surveying and special investigations (Table 6.2).

Table 6.1 Essential equipment and instrumentation for the geomorphological survey

Equipment and tools	Use
Topographic maps	Orientation on the ground and the geomorphological mapping of elements encountered during the survey
Geological maps and bibliographic materials	The geological and geomorphological mapping of the area to be investigated
Field records	Write down the field observations, draw diagrams, sketch drawings, bring the numbers and locations of samples collected and everything that cannot be represented on map
Stationery (pens, pencils, colored pencils, erasers, marker pens, etc.)	Writing, drawing and mapping
Binoculars	Perform remote observations, including places nearby that are inaccessible
Portable shears	Proceed through any scrubland and undergrowth and contribute to the cleaning of the outcrop
Trowel or similar tool	Clean the outcrops
Geologist's hammer	Detach rock samples from the substrate and observe the inner surface. Break pebbles, conglomerates or other coherent rocks (breccia, travertine, limestone, volcanic material, etc.)
Chisel	Take samples from coherent rocks (travertine, conglomerates, quartzarenite, lava, etc.)
Collapsible measuring stick – 2 m	Take moderate measurements
Clinometer compass	Measure the arrangement of the layers, of bodies and sedimentary structures, of erosive surfaces, imbricated pebbles, tectonic elements, etc.
Non-magnetic lamina	Facilitate the measurements with the compass

(continued)

Table 6.1 (continued)

Equipment and tools	Use
Altimeter	Measure the altitude above sea level
Handheld GPS (Global Positioning System)	Automatically detect the geographical coordinates of a point on the ground
Portable laser distance meter	Facilitate width measure escarpments, the width of ravines, etc.
Digital still camera	Photographing landscapes, landforms, deposits, etc.
Hydrochloric acid diluted to 10 %	Ascertain the carbonate composition of clastic sediments and the possible elements of a cement that binds them
Munsell color tables	Evaluate the colors of pedogenised horizons
Magnifying glass (10×)	Observe the mineral composition, structure, texture and any fossil content of the sample
Granulometry comparator	Conduct a field evaluation of the size of grains and minerals that make up a rock or a loose sediment
Resistant plastic bags of various sizes	Collect samples
Thick tip permanent markers	Write on bags, the plastic labels and rock samples
Metallic ties	Close plastic bags
Plastic labels	Report the sample number if it is made up of loose sediment. The plastic label is inserted directly into the bag with the sample number facing outward in contact with the plastic
Sample bag	Have on hand the stationery and all that is frequently used in surveying
First aid kit	Primary and summary medication in case of accident
Flashlight	Explore small caves and/or use as a light source in case there is a need to remain in the country for prolonged periods
Backpack	As well as all the necessary material for the survey, food and first aid equipment etc., should be carried

Modified from D'Orefice and Graciotti (2015)

6.3.3 Synthesis and Processing of Field Data

At the end of the systematic surveying of land it is advisable to arrange the individual field documents (originals from the survey), reinterpret them and integrate them, also with the help of aerial photographs, and report them, in a clear and readable form, on a unique topographic base. This will be the basic document for the implementation of our final geomorphological map.

The geomorphological map is an element that should always be attached to the final technical report. The latter must also contain the previously collected documentary data in the published preliminary research, data taken from geognostic tests and laboratory analyzes and always an appropriate number of photographs, accompanied by clear and explanatory captions. The report must be written in simple and clear technical language that is easily accessible even for lay persons, as far as possible.

Table 6.2 Equipment and instrumentation necessary for particular purposes

Equipment and tools	Use
Foldable shovel	To perform small-scale excavations
Shovel	When you have to move higher volumes of earth
Pickaxe	Perform excavation and surface trenches in incoherent sediments
Clubs of various sizes	Break very coherent rock (for example, igneous and metamorphic rock)
Billhook	Facilitate the crossing of densely vegetated areas and cut intertwined shrubs of the outcrop
Protective helmet	Protect the head from falling debris from rock walls or loose deposits
Metric tape	Measuring elements of several meters in length
Graduated rods and poles	Measure the geometric variations over time of a morphological element
Manual auger	Conduct an expeditious investigation of loose sediment in the first subsurface. It is a light tool that, under optimum conditions, can reach 8–10 m in depth with sampling (every 20 cm) of disturbed samples
Motor auger	Expeditious core drilling of the ground up to 10 m depth for the taking of continuous samples in cohesive and granular soils, also saturated and below the groundwater level

Modified from D'Orefice and Graciotti (2015)

6.4   Determining the Activity State of Landforms

6.4.1   Problematic Issues

One of the most important tasks facing the geomorphologist during survey operations is determining the activity state of a form. This is a very sensitive issue, therefore, must be addressed in all its aspects with due care and professional integrity, without coming to any hasty conclusions. Very often, as we will see later, when it has to do with forensic investigations involving landslides, the geomorphologist is required to provide precise information about their state of activity. In fact, the effects of these phenomena on human activities are often a cause of long and troublesome legal disputes.

It is important to point out that it is not immediate, and often even impossible, to establish, during routine survey operations, the activity status of a form and deposits associated with it. In fact, the activity of a natural phenomenon can manifest itself in different ways, making it:

(a) directly perceptible to the human eye (direct perception);
(b) recordable only by instrumental means;
(c) able to be reconstructed through comparative studies and analysis that consider the changes of form for long periods of time, or be hypothesized based on theoretical arguments.

Direct Perception The activity leaves evidence identifiable through direct ground observations. These are phenomena that evolve at a moderate to extremely rapid speed, the effects of which are perceptible to the human eye. Included in this category are some types of landslides (fall, topple, debris flow, debris avalanche, as well as slides and flows, characterized by kinematics fast enough to be perceived by humans), avalanches, volcanic outpourings and some forms related to the river and coastal erosion.

Instrumental Recording The phenomena are characterized by a speed of evolution from slow to very slow, such that it is not felt by people during the land surveying. However, as they are detectable changes in a relatively short time, they are measurable only instrumentally using programmed monitoring campaigns. Some gravitational landforms (sliding, flows, solifluction), badland and cryo-turbation landforms, rock glaciers, many wind landforms (Fig. 6.3) and most of the morphological elements produced by erosion or by river and coastal sedimentation may be included in this group.

Reconstruction by Means of Studies, Analyzes and Hypotheses These are phenomena that evolve very slowly (therefore are not perceptible to the human eye), and over a very long time. Therefore, it is almost never appropriate, temporal and economic level, perform geotechnical monitoring to measure their evolution, especially in the context of a forensic investigation to be completed relatively quickly. Phenomena that evolve very slowly (deep-seated gravitational slope deformations, shapes originated by karst, etc.) or intermittently (some phenomena related to volcanic and tectonic processes) usually fall into this category. Comparative reconstructions make use of the comparison between the morphological context recorded

Fig. 6.3 Stretch of the dune cord partially fixed by vegetation (Lacona, Elba island, Italy)

at the time of surveying and that detectable by documents, iconography, or photographs (aerial or terrestrial), produced at different times. The comparison between past and present will determine if the form currently detected was absent, or less evolved, in the earliest records. In this way it is possible to estimate, or even measure, the overall size of its evolution (for example, the retraction of a beach or a cliff, the changes of a relief affected by a deep-seated gravitational slope deformations, the movement of a rock glacier, etc.). Studies based on the influence that the evolution of a form has on human works, such as lesions produced in artifacts due, above all, of landslides are also very important. Useful information on morphological development in a specific area can be also acquired by means of targeted interviews with residents who, at times, represent a real historical memory of the physical knowledge of a given territory.

In general terms and for simplification, the main activity states of landforms and associated deposits and their definitions are summarized in Table 6.3.

6.4.2   Determination of Activity States Through Qualitative Analysis

Usually, during field activities, the geomorphologist must assess directly in situ the space-time evolution of individual landforms (for example, the retraction of a river embankment, the extension of a sinkhole (Fig. 6.4), the variations in height of a topographical surface subject to the spread of washout processes, etc.) also by means of expeditious measures to be performed in a short time or, in any case,

Table 6.3 Definition of the activity states of landforms and associated deposits

Activity states	Description
Active	Landforms (and associated deposits) that show evidence of activity at the time of surveying, whether it be a continuous, intermittent or variable activity phenomena
	Landforms (and associated deposits) which, although not active during the survey, usually manifest their activities during the year (seasonal activities) or at short time intervals (frequent activity)
Dormant	Landforms (and associated deposits) that do not show evidence of activity at the time of surveying, but for which it is possible to hypothesise a reactivation by the process that generated them, without this being necessary to invoke significant changes in morphoclimatic conditions
Inactive	Landforms (and associated deposits) that have no reasonable possibility of reactivation, as generated in morphoclimatic conditions that differ greatly from existing ones, or because subsequent topographic changes to the area do not allow it
	Unique activation phenomena (for example, rockfalls), although occurring in the course of surveying, also fall into this category

Modified from Dramis and Bisci (1998)

Fig. 6.4 Sinkhole pit collapse with vertical walls (Camposecco, Simbruini Mounts, Lazio, Italy)

compatible with those provided by the forensic investigation. These measurements can be performed with the tools that are commonly part of the geomorphologist's field equipment (see Tables 6.1 and 6.2). Interesting information about the extent and occurrence of geomorphological phenomena in the study area may result from interviews with the local population.

Other data, including a longer observation period, can be derived by an examination of existing documentation elements (maps, prints, paintings, photographs, reports, minutes, etc.) as well as by photointerpretative analysis of a multi-temporal type. But it is above all thanks to his surveying experience that the geomorphologist, operating on the ground, grasps the clues and those specific elements that contribute to providing a first qualitative evaluation of the state of activity of a particular phenomenon. In this context, it is important first of all to examine the orographic, hydrographic, lithological, structural and geomorphological framework of the area in which the form lies, from which the state of activity must be established.

In a second step, we should direct our investigation towards a series of observations, considerations and assessments, concerning, more closely, the form in question (and the deposits associated with it). The issues may be, for example: slope breaks, carvings, steps, undulations, swelling, subsidence, counter-slopes, fractures, partial or total coatings by debris or eluvio-colluvial products, injuries and changes in structure of human works, etc. Within this area, it is also important to assess the degree of surface alteration of the landform or of the associated deposits, the development of pedogenetic cover and of the vegetable on it, the remodeling amount of the form itself as a result of processes that are different from genetic ones. The analysis of these elements can often provide an indication of the timing of the last activation of a form in its present dormant (or inactive) state, and in some cases outline space-time evolution, at least for the most recent phases.

Fig. 6.5 Misalignment, with regard to the verticality of a pole of the power line, due to a landslide (Maierato, Vibo Valentia, Calabria, Italy)

Among field observations of a geomorphological type, the misalignment evaluation, with regard to the verticality of poles to support electric lines (Fig. 6.5) or telephone and palisades made of materials of various kind for delimiting private property or state property is particularly important. In this case the simple observation of the misalignment, excluding a priori any human interventions (in cultivated fields misalignment is often due to the stresses induced on the poles from agricultural machinery), can provide useful information on the presence of ground surface movements on which is based the work, primarily related to gravitational dynamics of slope or fluvial-denudational erosive phenomena. Their degree of misalignment and tilting can be adopted for a very qualitative estimate of the intensity of morphogenetic processes operating in the area. Considering the time interval elapsed between the start of works (placement of electric lines, fences, etc.) and the time of surveying, it is also possible to empirically estimate the speed of the phenomenon, assuming that the work has been done so correctly with regard to the verticality and the alignment, that the phenomenon is not very activated too long a time after its installation and that it has not suffered significant periods of stasis.

Also from examination of the tree and vegetation present along wooded roads made in historic times and in wooded slopes important considerations may be derived. In fact, some morphological characteristics of trunks, such as abnormal curvatures (hook-shaped, typical "S" shapes, etc.) (Fig. 6.6), prominent inclinations upstream or downstream in relation to their natural position (Fig. 6.7), impact scars

Fig. 6.6 Solifluction phenomena, highlighted by sloped and hook-shaped tree trunks, that have affected the altered surface cover of the outcropping pyroclastic deposits at the foot of Colle Volubrella (Municipality of Rocca di Botte, Abruzzi, Italy)

Fig. 6.7 Shrub vegetation shifted and counter-slope tilted following a sliding surface landslide (City of Rocca Santo Stefano, Lazio, Italy)

visible on bark, partial burial of their base or damage to root systems can provide valuable information indirectly on the state of activity of gravitational sloped forms (solifluction, soilcreep, shallow landslides, debris flow, etc.), or river forms and slopes due to washout (fluvial erosion of river banks, fans of flooding, concentrated runoff furrows, areas in ravines, etc.).

More data on the state of activity of a landform and the intensity of the phenomenon can be derived from the analysis of lesions or variations in anthropogenic structures such as roads (Fig. 6.8) buildings, viaducts, retaining walls, tunnels, etc. Of course, these data should be taken with extreme caution. In fact, in order to avoid erroneous considerations, it is essential to know the history of the structure that you are considering. A fair assessment of the state of activity can only be performed if the work has been designed and carried out properly, excluding, therefore, that any lesions, sagging, changes in structure, collapse etc., are to be attributed to structural defects in the work itself.

To determine the activity status of a landslide, in addition to the evidence mentioned above, during the survey particular morphological elements such as detachment niches, engravings, steps (Fig. 6.9), scarps, waves, sinking, counterslopes, lateral cracks, fractures, irregularities in the pattern of drainage, water stagnation areas etc., should be focused on. The qualitative evaluation of the "freshness" of

Fig. 6.8 Obvious fracture on a paved road, a few centimeters in width, generated by a landslide (Umbria-Marche Apennines, Marche, Italy)

Fig. 6.9 Turf cut sharply by a step generated by a surface flow landslide (City of Fara in Sabina, Lazio, Italy)

these findings can help to provide indications as to their evolution, such as the presence of steps with sharp edge dissecting clearly the turf, deep gaping fractures even some meters long and not yet filled by secondary material, surfaces with fresh fracture zones and *striae*, subvertical cliffs not yet altered, colonized by vegetation and affected by surface erosion.

In general, the presence of a reshaping of a form by morphogenetic processes different from those that have generated it, of a surface alteration, of a plant cover and the absence, at the time of surveying, of the necessary conditions for its reactivation, are indicators that help shape its likely state of inactivity. In this context, we can cite as an example: a landslide located on a valley floor and a significant distance from the slope that generated it, now completely covered by a forest of oak trees; a moraine cord of Late Glacial Age characterized by a rounded top, from the sides not very steep, from an area colonized by vegetation and affected at the base by colluvial products, often pedogenized; a rock glacier, with a face that is not very steep, stabilized boulders colonized by lichen and remodeled by surface runoff processes.

It is important to point out that to avoid confusion and misunderstandings, in the technical report the geomorphologist should fully disclose the criteria used in the evaluation of the state of activity of the examined landforms. It is therefore always best to clarify that the distinction between active, dormant and inactive landforms was only made on the basis of field observations of an essentially morphological type, alongside photointerpretative analysis, without the aid of targeted imaging techniques. In this case, therefore, the attribution of activity states, being qualitative, gives a purely indicative significance, and should be considered as such.

6.4.3 Definition of the State of Activity by Quantitative Analysis

When the geomorphological survey is aimed at forensic investigations concerning the assessment of geomorphological hazards linked to the spatial-temporal evolution of a landslide or avalanche, the retreat of the coastline, the degree of subsidence in a specific area, the topographical variation slope, or the estimation of the extent of surface erosion etc., it is essential to operate with appropriate methodologies and technical instruments. This is because, as pointed out previously, the erosion and accumulation landforms originating from different morphogenetic processes are manifested with speed, intensity, and with very variable return times.

The speed of evolution of a certain landform, such as a landslide, can be very slow and continuous (continuous activity) in a given time interval (mm/year or cm/year). Sometimes, however, the speed can be very fast (m/s) and intermittent, with "shots" alternating with long periods of quiescence, even lasting years (intermittent activity); at other times, it may be continuous but variable in time, with periods of acceleration alternated with periods of deceleration (variable activity).

Landforms related to the subsidence phenomenon can affect a given territory in a homogeneous manner, with a very slow and uniform speed (mm/year) even for long periods of observation. In other cases, they are subject to sudden acceleration, with high subsidence values, concentrated into short periods of time (dm/year even m/year). Some typical erosion forms of the morphogenetic fluvial-denudational process, such as the breaking of a natural river embankment, bank erosion (Fig. 6.10), etc., may occur suddenly with high intensity and with short and regular turnaround times. Other times the same phenomena may present medium-term recurrence times (between 10 and 100 years). Finally, there are landforms which are not characterized by signs of evolution in the current meteorological conditions, because they were formed during previous climatic phases; as an example, we can mention coastal wind dunes and moraine deposits belonging to the last glacial period.

From the above, it follows that the measurement of the displacement vector of some geomorphological phenomena is complex and not always feasible, and requires the use of suitable technical equipment. Only through the geotechnical monitoring it is, in fact, possible to reconstruct the evolution of some landforms in a given spatial-temporal context, with high accuracy (to the millimeter). This activity, which requires not indifferent economic availability, however, has meaning only if protracted for a sufficiently long period of time, of the order of several years.

The choice of instrumentation to be employed depends upon several factors. It depends on the type of process to be examined, the extent and location of the area and the accuracy to be obtained, but mainly depends on the time and the financial resources available for forensic investigation completion. As stated earlier, the monitoring has validity only if carried out for significant time periods, which often do not correspond to those granted by the judicial authority.

Some tools, such as strain gauges, distometers or fissurometers, accurately measure the performance of surface discontinuities, by thoroughly monitoring the

Fig. 6.10 Bank side
erosion of Cammarano
ditch caused by the
grinding work of the
riverbed (Carsoli, Abruzzo,
Italy)

micro-movements that can be recorded in the presence of fractures, cracks and discontinuity, found on the ground or in man-made works. Other devices, such as multipoint extensometers and inclinometers, are used to asses soil displacement in
depth, such as the sliding surface of a landslide. Other intervention methods are
used for the monitoring of geomorphological phenomena of a certain size (laser
radar technology, topographic surveying and GPS survey). These methodologies
and survey techniques, considered by the authors the most commonly used (but not
exclusive) for proper evaluation of geomorphological evolution of a given area, can
be appropriately integrated with each other to get results that are more reliable and
detailed.

   Below are summaries of the most commonly used tools and methods of investigation in the geomorphological field, particularly with regard to probe inclinometers, of unquestionable usefulness in the deep monitoring of landslides.

6.4.3.1 Fissurometers

These tools, which are easy to use, manually measure the relative displacement between two fixed points on the sides of small cracks present on structures, masonry and/or micro fractures affecting rocky walls. The fissurometer is composed of two overlapping plates: the first is transparent and engraved with a reference cross, the second is carved with a millimeter grid. The two plates are fixed at the turn of the gap to be measured by means of special two-component adhesives or stainless screws. During installation, the center of the cross reference is made to coincide with the center of the axes of the Cartesian grid graph (Fig. 6.11). Once the date of installation has been registered, periodic measurements are performed, to be carried out according to predetermined time intervals. Any movement of the two flaps of the discontinuity are recorded on the graph grid according to the two x and y components. There are several types of fissurometer: linear, electrical, measuring single axis with mechanical comparator. Generally, the precision obtained in the measurements is of the order of a millimetre.

Fig. 6.11 Fissurometer composed of two superimposed plates in plastic material, with reading cross and measuring grid. The tool was driven by stainless screws, riding on a fracture which runs through pyroclastic rocks (City of Rome, Italy)

6.4.3.2 Strain Gauges

These instruments measure the relative displacement (one-two-three-way) between two fixed points on the sides of a fracture in the rock and made integral with the ground through appropriate anchoring systems. There are bar strain gauges for discontinuity measurements of the order of 1–2 m and wire strain gauges to size even greater fractures, up to tens of meters. The measuring principle is based on the relative displacement of one of the two edges of the fracture, where the extension of the bar or the expansion of the wire is originated; a special sensor registers the movement and transmits it to the control unit. The accuracy that is achieved with these devices is of the order of a tenth of a millimeter. Strain gauges are particularly suitable for the measurement and monitoring of surface morphological elements such as fractures, trenches and cracks present along slopes affected by deep-seated gravitational slope deformation or landslides in rock masses.

6.4.3.3 Multipoint or Depth Extensometers

These are tools used to check for any ground and/or rock displacement, at various depths in the subsurface, in most measuring points (multipoint are defined when the points are more than one), located within the boreholes. The equipment is composed of invar cables or fibreglass rods, fixed inside boreholes at predetermined depths. The surface wires are wound around a pulley, that can rotate around a recording potentiometer, and are tautened with weights. Any displacements of masses of soil to a certain depth, such as along a sliding surface of a landslide, directly modify the length of the cable placed below such level, leaving stable those placed at shallower depths. These cable tension changes are automatically recorded, with an accuracy of 1 mm.

6.4.3.4 Distometers

Very similar to strain gages, these manually measure the changes in existing distance between two fixed points (linear movement), placed on either side of a discontinuity with a considerable opening size. The distometer consists of a cylindrical body made of light alloy, a tape invar alloy, a digital dial gauge, a tensioning system. There are two types of distometers: wire and strip iron. In the first, the basics of measurement can also reach a distance of 50 m; in the second, bases have a slightly shorter distance. The accuracy that is achieved with distometers is very high, approximately 0.02 mm for measuring lengths of up to 20 m. As with strain gauges, distometers are widely used in the measurement of surface dislocations of the slopes affected by gravitational processes.

6.4.3.5   Probe Inclinometers

Probe inclinometers are composed of an inclinometer tube, an inclinometer probe, an electronic control unit for measuring and recording, and accessories such as cables, pulleys, winding drum, etc.

The inclinometer tube is made of deformable material and/or flexible aluminum or PVC (vinyl chloride polymer) to follow any pressures arising from the movements of the ground at different depths. The tube is driven into a previously created borehole, and suitably cemented so as to be integral to the ground that encompasses it. The cementing of the pipe must take place at low pressure with cementitious mixture consisting of water, pozzolana cement and bentonite, mixed together in various proportions (generally using a ratio equal to 100, 30, 5 parts by weight respectively). The tube has a circular section of approximately 76 mm of inner diameter, provided with four vertical grooves, perpendicular to each other, along which slide the wheels of the cart of the probe during the measurement phases. The inclinometer pipes are usually made of clips from 3 m in length each and are assembled by means of suitable coupling sleeves. Special care must be given to their installation. An incorrect installation of the tubes may, in fact, affect subsequent measuring operations. In summary, the following requirements and precautions should be taken: the perpendicularity of the tube must have a permissible tolerance of about 1–2°; coiling must be less than 0.5°/linear meter; the bottom of the tube has to be closed to the concrete cap and suitably anchored to the stable substrate for a sufficient depth; the orthogonal grooves (guides of the probe) must be free from obstacles and steps that may form in the joints; finally, the tube head, in its part emerging from the ground level, should be appropriately protected with a suitable manhole with a lock, to prevent accidental damage and vandalism.

The inclinometer probe consists of a cylindrical steel casing, mounted on small flexible wheels and tautened by springs. In its interior, it is equipped with a high accuracy sensor (servo-accelerometer) which, urged by the force of gravity, generates an electrical signal sent by cable to the control unit. The probe measures the angle (inclination) formed between its axis and the vertical in the plane containing the wheels. Measurements are performed at fixed points (generally with intervals of 1 m or 0.5 m) during the ascent phase of the probe from the bottom of the hole, where before it had been lowered. These are repeated, rotating the probe 90°, to minimize any systematic errors. At the end of the measurements both the form of the displacement vector at the various depths investigated (local differential from the bottom) and the top of the inclinometer tube (integral differential from the bottom), as well as its azimuth, i.e., the angle formed by the resultant of the displacement and the geographical east (positive counterclockwise) are obtained.

With the same probe, several appropriately installed inclinometer tubes in the area to be investigated can be read. These are to be read a first time, said "zero reading", which will be compared with subsequent measurements, carried out according to regular, predefined time intervals.

The data acquired on each inclinometer are restored according to graphs that have on the abscissa the result of the displacement expressed in millimeters and on

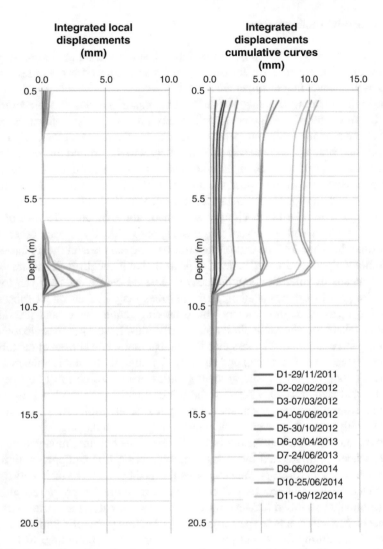

Fig. 6.12 Graph of reading of an inclinometer tube, at specified intervals. You may notice local and integrated movements to single depth (left panel) and integrated displacements accumulated in the pipe head (right panel). Each curve is identified by a specific color according to the date of the measurement. The displacements, concentrated in the range of 8–10 m, with the higher values at about 9.5 m deep from the ground level, correspond to the sliding plane position of a landslide (Source: LM Puzzilli)

the ordinate, the depth of the inclinometer tube in meters. Each curve, characterized by a specific color, identifies the reading of the measures at a given date (Fig. 6.12). Next to the curves of the local and integral displacement, azimuth curves are also represented, which differ from the first only in the abscissa values that, in this case, are expressed in degrees from 0 to 360°.

It is clear from what has been described above that probe inclinometers are very useful in the field of geomorphology, especially for the deep monitoring of landslides, for the assessment of the state of activity of a landslide, and for the developmental study of a slope. The data obtained are very precise; in reading graphs, even millimetric displacements are noteworthy. The deformations undergone by the soil, in the orthogonal direction to the axis of the borehole, can be determined at various depths, according to the "step" that is adopted for manual lifting of the inclinometer probe. It follows that it is also possible to detect multiple sliding surfaces.

In view of their high accuracy, probe inclinometers are particularly suitable for the monitoring of landslides characterized by slow to very slow shifts. When the monitoring is continued for sufficiently long periods, it is possible to determine, in addition to the depth of the surface or of the sliding surfaces, also the speed, the direction and the magnitude of the vector displacement of the landslide. Useful information on the characteristics of the landslide are provided indirectly even when the inclinometer tube undergoes significant deformation, such as not to allow the passage of the inclinometer probe up to the bottom of the hole. In this case the abrupt and sudden interruption indicates a sudden reactivation of the landslide movement at a specific depth, coinciding with a sliding surface, a movement that manifests itself with such intensity as to cause the rupture of the inclinometer tube, which, it shall be recalled, is aluminum or PVC.

6.4.3.6   Laser Surveying Technique

The first measurement devices that use laser technology were created in the late 1970s. The principle on which this method is based is that of measurement, by means of a laser beam, of the distance between the emitter means and the object. This measurement may be performed with either the pulsed method, which calculates the travel time of the wave between the emitter and the object, or by the phase method, which calculates the phase shift between the wave transmitted and return wave.

This method of surveying shows significant potential both for the possibility of acquiring remote data and for high speed acquisition (thousands of points per second).

Laser tools can be divided into two main categories: terrestrial laser scanning, TLS (*Terrestrial Laser Scanner*) and laser scanning from either satellite or aircraft, LiDAR (*Light Detection and Ranging*). Both of these instruments allow the planoaltimetric position of a considerable number of ground points to be determined with great precision and in very short times, without the aid of tools reflectors.

TLS technology allows very high accuracy to be achieved, typically of the order of millimeters, but its use is a function of the distance between the tool and the object to be monitored, as well as weather conditions. Depending on the equipment used different levels of accuracy may be achieved. With a pulse tool, located 100 m away from the object to be monitored, accuracies varying between 2 and 5 mm are obtained. The maximum measurable distance is about 1,000 m. This technique is suitable for the precise and detailed reconstruction of slope morphology and consequently, also for the definition of the main geomorphological instability phenomena that occur on them.

LiDAR instrumentation is more complex because the system, usually installed on aircraft or helicopters, is composed of different units. The aerial platform has a laser sensor that sends a beam of waves in the infrared frequency (transmitter) and collects the return echoes (receiver). A system for GPS satellite positioning and inertial navigation system (INS) provide constant guidance and allow the routing of aircraft employed. LiDAR technology allows fast data acquisition with an accuracy of a few centimeters over large areas. This methodology, after suitable filtering, is very useful for the realization of digital terrain models (DTM) to high resolution and geo-referenced, to be used in monitoring particular areas in evolution, subject to instability and/or deformation, namely: volcanic areas, slopes affected by gravitational dynamics, coastal areas subject to marine ingression, floodplains exposed to hydraulic risk, areas subject to mining activity.

Factors limiting the use of laser technology are represented by the thick vegetation and in the case of LiDAR, slopes characterized by sub-vertical walls.

The laser technique can be integrated with the overlap of photogrammetric digital images; so you can get the digital terrain models which can be very similar to reality.

6.4.3.7 Radar Technology

This remote sensing technique, developed in the early 1990s for the detection of surface movements of the ground, is based on the propagation of electromagnetic waves in the microwave range and on the detection of the relative diffusions. This survey is particularly used in the field of geomorphology because it is not affected by weather-climatic conditions (cloud cover) and by lighting (day or night). The instrumentation can also be installed on satellites in orbit around the earth, on aircraft or on terrestrial equipment. The radar device is constituted by some main elements: the transmitter that emits the electromagnetic waves; the antenna that sends waves to the target (exposed rocks, hard non-vegetate surfaces, large boulders, buildings, pylons, guard rail of roads, and viaducts, etc.); the receiver that picks up the reflected wave from the target; a recorder which stores the signals that come back to the source.

Based on the characteristics of the return signal (distance, intensity, and phase), it is possible to obtain the properties and location of the object detected. At the end of radar prospecting a two-dimensional image of the earth's surface investigated is obtained, which will be distinguished by different shades of grey depending on the recorded signal power.

To improve the resolution of the radar survey you can resort to so-called SAR technique (*Synthetic Aperture Radar*), which consists of increasing the measurement time of an object in proportion to its distance from the sensor. The radar in SAR methodology is mounted on satellite or aircraft and directed towards the earth's surface, in a position orthogonal to the direction of its movement. Two images taken in different times of the same area can be compared using the InSAR technique (*Interferometric Synthetic Aperture Radar*). By means of suitable processing systems, at the end of the relief, it is possible to estimate soil displacements with centimetric accuracy according to the phase shift of the reflected waves during the two consecutive acquisitions of the same object. When a given area is studied, special artificial reflectors that are easily identifiable on the radar image can be installed to enhance the power of reflectivity.

With the SAR technique, it is also possible to monitor large areas, detecting ground motion with high accuracy. This type of relief is therefore widely used in the geomorphological field, mainly for subsidence studies, to define the evolution of the areas affected by deep-seated gravitational slope deformations, by landslides or by intense denudational erosive processes, to analyze the development of the river dynamics in areas prone to flooding and lateral riverbank erosion.

When radar equipment is placed on the ground (*Terrestrial InSAR*) it is possible to carry out a larger number of scans of the same area, increasing the speed of execution and the precision of the relief. In this case, however, the instrumentation must be placed at a maximum distance of 4 km from the subject to be observed. This condition, in certain circumstances, may pose a serious limitation to its use. At the end of the survey, by processing and comparing the series of images obtained, it is possible to determine the displacement vector of the land with an accuracy of less than 1 mm.

6.4.3.8   Geodetic-Topographic Survey

The measurement techniques in geodetic-surveying have been performed with theodolites, total stations (motorized and automated) (Fig. 6.13), infra-red and laser distance meters, electronic levels, etc. They allow the plano-altimetric position of representative points of the earth's surface to be obtained, with considerable accuracy. The points of which you want to determine the three-dimensional coordinates are materialized by means of special irremovable pillars, effectively founded in the ground, which constitute the so-called network of target points. This type of survey can be used both to represent the exact geometric position of certain morphological elements, or for areas that are particularly interesting from a geomorphological point of view, both for the monitoring of unstable areas, such as areas in landslide or subsidence.

For measuring, at predetermined intervals, the displacement vector of points, suitably chosen and materialized on the ground, you need to find stable areas, far removed from the area being studied, where the reference strongholds may be placed. The latter will enable the development of the network of measuring target points that will be observed periodically with relevant technical specifications. The first set of topographic measurements on all points of the network constitutes the "base reading" or "zero reading." The subsequent series of measurements, carried out according to a predetermined scanning time, compared with the zero reading will allow the detection, with a precision of a millimeter, of the possible displacement vector relative to the individual points. The precision achieved by the monitoring activities depends on the instrumentation used, and the geodetic-surveying

Fig. 6.13 Total automatic station with laser distance meter (Source: D. Matarazzo)

technique used (triangular-trilateration, traversing, celerimetric survey), but mainly depends on the correct design and materialization of the topographical network of cornerstones, since the points must be visible to one another and the network of points must have a special geometrical shape with sides and angles distributed with a certain uniformity. As previously mentioned, the reference target points must be located in stable areas; that is, those that are not affected by the movement of the terrain in the three components (x, y, z). This must therefore be done with extreme accuracy, possibly with the assistance of a geomorphologist to avoid areas prone to instability phenomena.

6.4.3.9 GPS Survey

GPS (*Global Positioning System*) is a global positioning system of points on the earth's surface based on the reception of radio signals from satellites orbiting Earth. The set of satellites forms a constellation. This system, born in 1973 in the United States for military purposes, is now widely used for civil applications. In addition to the American constellation there exist a Russian satellite system (*GLONASS*), a European one (GALILEO) and a Chinese one (*COMPASS*). The set of different satellite positioning systems takes the name of *Global Navigation Satellite System* (GNSS).

The GPS survey technique is based on the use of special tools (receivers) which pick up radio signals from satellites (transmitters) and with rapid calculations determine the exact position of the observer (coordinates x, y and z) in an earth-centered reference system fixed relative to the earth. This system was established in 1984 as *World Geodetic System* (WGS84). Each GPS receiver consists of an antenna which

Fig. 6.14 GPS receiver, (Municipality of Posta, Lazio, Italy) (Source: D. Matarazzo)

picks up the radio signal emitted by the satellites (Fig. 6.14), a received data processor, a display for viewing information, a connection system with other receivers to perform special precise measurement techniques.

The use of the GPS technique compared to traditional geodetic-surveying methods offers multiple advantages:

- the measurements are not affected by weather and climate conditions;
- precision of the method is the same in all the points that have been measured;
- the reference system is homogeneous;
- the chosen points to perform the control measures and monitoring of a particular phenomenon must not be visible to each other;
- the recognition of land needed to properly design a network of target points to be used as monitoring, is simpler and faster compared to the design of a measuring network using topographical methods.

The only limitation of this system is that of poor or lack of reception of the radio signal with reference to urban centers and dense vegetation.

There are three different types of GPS positioning:

1. *Absolute positioning*. The coordinates are provided by a single receiver that acquires the GNSS signal code. The data is available in real time but the precision obtained in positioning is in the order of several meters (10–15 m). The receivers are easy to use and small size (pocket or handheld).
2. *Relative positioning*. The coordinates are given by two receivers, kept in a fixed position for a time ranging from minutes to hours, that acquire the code and the

GNSS signal phase. The data is available after a process of elaboration and are therefore not in real time, while accuracy is reached in positioning is in the order of centimeters. The receivers are mounted on tripod or graduated rods.

3. *Differential positioning*. You use a receiver that acquires the GNSS signal code and the differential corrections relating to the code of another receiver in a known point. The data is supplied in real time with an accuracy in the placement of about 1 m.

The GPS measurement techniques can be traced essentially to two categories: static and kinematic.

The first category can be divided in turn into static and rapid static relief. The second continuous relief, *real time kinematic* (RTK) and Stop and Go.

Static measurements in the two receivers remain on the same spot for a very long period (of the order of hours). With this method, very high precision in the positioning of the points to be monitored, of the order of some millimeters, is obtained.

In kinematic measures a receiver, called *master* is placed on a base point, another receiver, called *rover*, moves on points of detail to be detected. As part of this technique, *real time kinematic* (RTK) relief is particularly used in the field of geomorphology. It allows the rapid acquisition of points (in a matter of seconds) and thanks to the differential correction it is possible to obtain accuracies to the centimeter.

In the field geomorphology, the GPS survey technique it is widely used in virtue of the costs of the receivers, which is not too high, the fact that it is easy to use and that satellite coverage is now guaranteed worldwide. This technique has particularly useful applications in the production of digital terrain models of the areas affected by evolving fluvial-denudational dynamics, monitoring of landslides and detailed detection of particular erosion and accumulation landforms.

6.5 The Detection of Surface Deposits

6.5.1 General

Geomorphological surveying, in addition to the study and interpretation of erosion and accumulation landforms, allows for the systematic analysis of both deposits associated with accumulation landforms and of those that are superficial, i.e., of those sediments that are not delimited by an external surface distinguished by its own configuration (see Sect. 6.2.2).

Their sedimentological, stratigraphic (particularly mineral-petrographic, paleontological, palynological, paleobotanical, archaeological), and structural characteristics in fact provide morphogenetic, paleoenvironmental, paleoclimatic, and chronological information.

This information, if supplemented by a detailed analysis of erosion, sedimentary and deformation episodes, recorded by significant sequences of surface sedimentary[1] bodies, as well as possibly outcrops along natural or man-made sections, can contribute to the reconstruction of the most recent evolutionary events of the examined territory.

While on the one hand, the study of surface deposits is obviously of scientific interest, with repercussions in geomorphological mapping, on the other hand, the same study is also of great importance under a geological application profile in the forensic field. In fact, most infrastructures (buildings, bridges, roads, viaducts, etc.) and human activities (quarries, landfills, etc.) interact directly with this type of sediment.

The surface deposits we usually face during our geomorphological investigations are largely of continental and coastal origin. All this makes it considerably difficult to detect them for a number of reasons, listed below, which are mainly related to their nature and the characteristics of the environmental conditions in their formation:

- frequently lateral stratigraphic discontinuities;
- abrupt and repeated lateral vertical variations in sedimentary *facies* even over short distances;
- no direct overlap of subsequent age sediments ("snap" and stepover);
- generally fragmented areal distribution;
- frequently moderate thickness;
- scarcity and fragility of the outcrops;
- shortage of useful elements for chronological or paleoenvironmental attributions;
- shortage of reference levels (*marker*).

For the reasons mentioned above, the detection of surface deposits requires, in addition to suitable training in geomorphology, also adequate knowledge of other disciplines, such as geology of surface formations, sedimentology of continental and coastal environments, volcanology, pedology, geochronology, etc.

6.5.2   Description of the Outcrops

Given the great importance of the superficial deposits in geomorphological analysis, it is essential that during field surveying they are described in a systematic way, both when they are associated with accumulation landforms and when they lack their own characteristic shape.

---

[1] Sedimentary body is defined as a portion of a local sedimentary sequence in which both the internal sedimentological features and the discontinuities that delimit it in space allow its differentiation with respect to other parts of the same sequence.

During their study is important to determine their genesis and reconstruct the geometry of sedimentary bodies, the paleogeography of the area and the evolution of geological events that have affected the examined territory.

During surveying, the deposits identified in outcrops of modest dimensions may certainly be described quickly and in good detail, while those exposed in broader sections (anthropogenic and/or natural), will require more time (Fig. 6.15). In these cases, the survey must be performed with more complete equipment (shovel, pick, metric tape, protective helmets, etc.) than that which is normally used (see Table 6.2).

If the outcrop is not very recent, and then was exposed for a time to the action of exogenous agents, it will present more or less vegetated, affected by coatings of surface alteration and possibly covered with earth which has fallen from above. These elements may deceive us and lead us to erroneous suppositions. In such cases it is necessary to thoroughly clean the surface of the exposed section with a trowel or spatula, or, in the most sensitive cases, with a brush. If the wall is very extensive, can be employed a shovel, the use of which can also be restricted to the most representative areas, such as the limits between the various sedimentary bodies. However, cleaning must be done carefully so as not to hide or destroy some sedimentary structures that are prominent on an altered surface, such as parallel or cross lamination in fine sediments, or lenses and levels concretioned by secondary circulation waters, which protrude by selective erosion.

After cleaning the outcrop, it is essential to perform the first observations from afar, to try to identify the general characteristics of the section, and then close up.

Fig. 6.15 Artificial section in lacustrine deposits of the Lower-Middle Pleistocene, characterized by alternating layers of silt clay and sandy silt. The layers are inclined by about 20° towards the SSW (to the right of the figure). Lacustrine deposits are also affected by an extensional tectonic dislocation, with throw of some meters in the eastern portion of the section (on the left of the figure) (Conca Intermontana of Oricola-Carsoli, Abruzzo, Italy)

The geometry of the main sedimentary bodies and some large-scale sedimentary structures (channels, erosive surfaces, dislocations, etc.) are identified better from a distance.

The best way to describe the individual elements that make up the outcrop (main sedimentary bodies, discontinuities, deformation structures, sets, layers, laminations, lenses, channels, etc.), their geometry, and their spatial relationships is to draw a stratigraphic section. To this end, in relation to the outcrop, a sketch or preliminary outline shall be made in the field book or other media, summarizing the fundamental characteristics in a clear and concise way that takes into account both the angular relationships between the various elements that constitute and the scale.

In this first schematic representation, where various specific observations can be recorded that were made in searching for and locating samples, it is important to also report the possible presence of a bedrock, of soils or paleosoils and of an eluvium-colluvial cover. It will also be advantageous to orientate the outline, accompany this with a reference scale and complete it with the addition of a date and any place names.

The schematic drawing should be necessarily supported by photos, which should be taken from a direction as orthogonal as possible to the plane which contains the exposed surface area and from different distances, so as to acquire images with different degrees of detail. In doing this it is important to add a dimensional reference (measuring rod, pick, hammer, meters, etc.) to the side of the section, in order to subsequently reconstruct the actual size of the subjects photographed.

In the field book, or other medium used, more information should be written down that is not represented in the graphic scheme, such as plano-altitude location of the outcrop, its nature (i.e. whether it is natural exposure, an artificial excavation, an anthropogenic embankment), its persistence over time (that is, whether it is a lasting or ephemeral exposure) and its size.

The graphic scheme of the outcrop, the annotations on the field book and the photographic documentation, constitute the essential elements to prepare the final stratigraphic section on the drawing board. On the latter, all the detailed observations carried out in the countryside and sampling points should be positioned with great precision. Once the general features of the outcrop have been described, you can go on to perform a close observation, which will involve a thorough examination of every detail. The examination should be conducted with a fairly critical approach, and as complete, consistent and objective as possible, avoiding the introduction of interpretative aspects during this operation.

At this stage, it is appropriate to follow a hierarchical criterion in the description, starting from the first main sedimentary body to then gradually move to any lower-ranking bodies (for example, a homogeneous set of concordant layers, a lens, a level, or even a single layer).

In any case, regardless of hierarchical order, in the analytical study of a deposit it is important to consider the following items for each identified body:

- thickness;
- geometry;

- grain size;
- sorting;
- texture;
- state of packing;
- consolidation;
- degree of cementation and nature of the cement;
- characteristics of the clasts and matrix;
- primary and secondary sedimentary structures;
- color;
- discontinuity surfaces delimiting the sedimentary body;
- tectonic and gravitational deformation structures.

The critical analysis of all the information listed above will allow hypotheses to be drawn up of any sedimentary facies recognized within the body examined and in the depositional environment.

Bearing in mind the specificity of the subject, please consult specialized texts, some of which are listed in the bibliography, for a more detailed description of the individual elements listed above.

To facilitate the collection of data on the ground and ensure their completeness, but also to standardize acquisition procedures, it is considered very useful to refer to the appropriate records for the detection of surface deposits. In this context records drawn up directly by the surveyor or those already published may be used as required.

All these elements will be subsequently included in the final report and will enable the judge and/or lawyers (in the case of expert witness or party) to provide detailed data that is easy to understand.

6.5.3 Instrumental Detection of Surface Deposits

Direct and indirect imaging techniques, complemented by laboratory analysis, must also be added to the field survey. The surveys will be carried out mainly in morphological fields conducive to sedimentation and preservation of surface deposits by erosion, such as coastal and river flood plains, river and glacial valley bottom, river, sea or anthropogenic terraces, karstic and tectonic depressions, orographic shelves, the piedmont and foothills bands, the basal portions of rocky cliffs, coastal bays.

When there are no natural or man-made sections in the area surveyed, which allow the study of surface deposits, at first it is necessary to retrieve all the data relating to the conditions of the subsurface of the area in question or its immediate vicinity, available from private or public entities (stratigraphy derived from mechanical surveys with continuous core drilling, water wells, excavations, trenches and earthworks of various kinds, penetration tests, geophysical prospecting). Only later, also based on the recovered data, you can plan a detailed campaign of geognostic surveys. This, according to the set objectives, may be conducted for further investigation, generally attributable to the following main types of investigation:

- expeditious survey in the first subsurface with light portable instrumentation;
- surveys in the first subsurface through excavations and trenches with a mechanical excavator;
- investigations by mechanical surveys;
- investigation by means of penetration tests;
- indirect surveys through geophysical prospection.

Within such types of investigations, the former will be described in greater detail because it is the one which is used most widely during geomorphological surveying on the basis of its simplicity of execution and ease of transport. Instead, regarding geophysical surveys, see Chap. 8 where the main methods used in forensic investigations are described.

Given the breadth and complexity of the topics listed above, for any further information, please refer to the numerous specialized texts in the literature.

6.5.4 Expeditious Investigations in the First Subsurface Using Lightweight Portable Instrumentation

By means of the excavation of small hand trenches and the execution of surface surveys with portable manual or power augers it is possible to explore the subsurface at a depth of some meters, in a short time and with low overheads.

6.5.4.1 Portable Manual Auger

The manual auger, commonly referred to as "Dutch auger", is an excellent investigation tool that, in case of scarcity of outcrops, enables the expeditious and effective exploration of the first subsoil. In fact, its light weight and easy handling make it easy to use even by a single operator. In particular, it allows the drilling of loose sediment that is tightly clustered, and generally not too coarse, up to a maximum depth of 8–10 m from the ground level. This performance is, however, dependent on the characteristics of the soil in which it operates.

The equipment consists of a tip with sharp edge, provided with a special shape to minimize friction during penetration and extraction from the soil, a handle and rods of 100 cm preferably with a "bayonet" attachment and a connecting socket (Fig. 6.16).

Using the handle the operator applies a rotational movement in the rods that transmit it to the tip. This being screwed into the ground for a short distance enables the removal of very disturbed cylindrical samples, of about 20 cm in length and 7–10 cm in diameter. The sample removed from the tip can be examined before the

Fig. 6.16 Portable manual auger with handle, connecting rods and tip with sharp edge

Fig. 6.17 Predominantly horizontally oriented drilling with manual auger in fluvial deposits (City of San Vittore del Lazio, Frosinone, Italy)

auger is introduced again into the drill hole for another extraction. Advancing by about 20 cm at a time you get a set of samples in sequence, that can be used for stratigraphic purposes and for laboratory analysis that, for the purposes of investigations, do not require undisturbed or material that can also tolerate alterations of the sample on a vertical interval of about 20 cm.

This auger can also be used on the bottom or on the walls of excavations or trenches (Fig. 6.17). Ultimately, it is a tool primarily designed to capture, quickly and with low costs, information on the thickness and nature of loose surface sediments.

6.5.4.2   Portable Motor Auger

With regard to portable power drills, there are various types available on the market. Among these, a model that uses core drilling technology with infiction with continuous coating of the core by means of a tubular sheath is particularly interesting. It is a light corer, designed for expeditious survey of up to 10 m in depth, which allows the taking of continuous samples in soils that are not too cohesive, from soft up to medium degree of consolidation, and granular, from well thickened to loose, also saturated and below the groundwater level. Such a device, initially successfully tested in the field of tephro-stratigraphy and stratigraphy of volcaniclastic products, has been used with satisfactory results, even in the context of lagoon, marsh, river, wind and beach deposits (Bellotti et al. 2011, 2016).

The apparatus is characterized by a contained volume, so as to permit, in the lighter configuration, the transport and the use of even a single operator. It essentially consists of a core barrel (drilling head, cartridge, the sheath, the connecting rods), an energiser (auto-motorised breaker) (Fig. 6.18), and an extractor (special mechanical binda lever). This type of corer uses innovative technology, which provides a coating of the core with a special protective tubular sheath, radial containment, at the same time as the advancement of the core barrel into the soil, thus avoiding the effects of rearrangement induced by friction that develop between the core and the inner walls of the core barrel. The corer, in fact, does not advance into the

Fig. 6.18 Light motor corer composed of the drill head, cartridge, sheath and connecting rods, containing the core covered in a protective sheath. The tool is driven into the ground by means of the thrust exerted by the auto-motorized breaker positioned at the top of the rods (Province of Caserta, Campania, Italy) (Source: M. Pantaloni)

Fig. 6.19 Continuous core extracted with light corer motor, ready to be sampled and analyzed in the laboratory after removal of the protective sheath

ground for a short distance, as in the most common mechanical drilling systems, but is driven in one sequence, until it reaches the maximum possible depth. Once extracted, it does not need any additional equipment to extrude the core from its container, thus also avoiding runoff problems to the sample from water used in this operation.

Thus, a continuous core diameter of the material of 38 mm, may be obtained, with a low degree of geotechnical disturbance, with 100 % recovery rates, on which to perform detailed stratigraphic, sedimentological, paleontological, palynological, pedological, etc., analysis (Fig. 6.19). The core covering sheath makes, moreover, the transport of the latter also easy in the absence of an appropriate container.

The coring technology shown here, in addition to allowing the crossing of loose sands and densified, silt, clay and peat without any difficulty (even in the presence of the water table), also allows for a high reading resolution of the stratigraphy of the samples – due to the low degree of disturbance to the core – in which the sedimentary microstructures may be preserved unaltered; for example, a thin plane-parallel and cross-lamination.

Ultimately this instrumentation is effective both as regards to the micro-stratigraphic detail, and in terms of operational practice. In the latter case, the limited overall dimensions and the manageability of the apparatus allow them to operate even in logistical conditions which may be particularly problematic for other traditional technologies (roadsides, wooded areas, intensively cultivated areas, etc.), as well as numerous samples of adequate length to be taken in relatively short times and with low costs.

6.5.5 Surveys in the First Subsurface Through Excavations and Trenches with a Mechanical Excavator

If you wish to examine the surface deposits of larger sections, to observe their lateral and vertical variations and the discontinuity of surfaces that define them, it is advisable to carry out minor excavation work on excavation pits or trenches with a mechanical excavator. The latter, if properly implemented and made safe, can reach a maximum depth of around 4–5 m and a width such as to allow easy access to the detector, also allowing the manual removal of any completely undisturbed samples that are cubic or parallelepiped shape.

The excavations and trenches are made after careful consideration of geomorphological and hydrogeological conditions of the area to avoid hazardous situations due to landslides and epi-superficial runoff and groundwater that might interact with the man-made work, compromising the integrity. In addition, before making excavations in the subsoil it must be ascertained whether there are any remnants of war and possible underground utilities (electric and telephone lines, gas pipes, water pipes, sewers, etc.).

Especially when working in the forensic field, to avoid disputes that could affect the investigations, the excavation work must first be authorized by the owner of the land and by the competent authorities and shall be conducted in full compliance with current health and safety regulations. In this context, it will be essential to put in place all technical operations necessary to ensure the stability of the trench walls. Finally, on completion of the investigation, the project area must be restored to its original condition.

6.5.6   Investigations with Mechanical Probing

Sometimes the geomorphologist in charge must operate on high consistency deposits (conglomerates, breccias, travertine, sandstones, volcanic tuffs, etc.), or must investigate the land at a greater depth. In these cases, it is necessary to use mechanical probes, using drilling equipment and drilling probes generally mounted on a truck or tracked vehicle. Depending on the purpose of the investigation, the type of terrain to be traversed, the depth to be reached and of the hole diameter, different drilling techniques will be adopted, which in principle can be traced to three major systems: percussion (with guide steel cable or hollow rods), rotational (hydraulic type or more rarely, mechanical type, i.e., with rotary table), and mixed, such as rotary and percussion drilling (i.e., with rotation of the percussion rods by means of a compressed air hammer placed at the hole bottom).

While percussion and rotary percussion systems do not lend themselves to the reconstruction of detailed stratigraphy of the subsoil, since the extracted material is gradually brought to the surface in fragments and chaotically (progress leads to the destruction of the nucleus), the rotating system is the most widespread (Fig. 6.20). This system also allows the recovery of continuous samples using appropriate core barrel equipped at the base of a sharp toothed crown. In this case, the samples may be considered to be disturbed if for their recovery a simple core barrel (hollow cylinder commonly with variable outer diameter of between 66 and 146 mm and a length between 500 and 3,000 mm), or be considered minimally disturbed if a double core barrel has been used (that is, a core barrel provided with a fixed inner cylinder and an external one that rotates rigidly with the battery of rods) which eliminates the problems resulting from the circulation of the drilling fluid and the torsion of the rotating outer cylinder. If a sample is needed with an even lower degree of disorder (for example, for the acquisition of geotechnical data), then so-called triple core barrels should be used, which allow the possibility of keeping the sample in suitable flat blank packages.

Fig. 6.20 Mechanical rotation drilling performed with crawler probe in the plain of Marina di Campo (Elba Island, Italy). The continuous core drilling was performed, along with five other surveys, in the context of the surveying of the geomorphological sheet at 1:50,000 scale "Elba Island" for the characterization of surface quaternary deposits

There is also the possibility of taking minimally disturbed samples in stretches of the survey that are deemed significant for the intended purposes, by means of infix-ion or percussion samplers. Among these, the simplest are those with thin walls, also termed *Shelby*, constituted by a die (galvanized steel cylinder between 600 and 900 mm in length, equipped with cutting shoe), which is driven into the ground by the weight of the battery of rods or, in rare cases, by means of a striking mass.

Mechanical surveys are particularly useful to geomorphologist working in the forensic field because, in addition to offering the opportunity to reconstruct an accurate subsurface stratigraphy of the area of operation and to take samples for laboratory analysis, they also permit the installation into the borehole of monitoring tools such as inclinometers, multipoint extensometers, piezometers, probes for temperature measurement, etc.

Inclinometers, as shown in Sect. 6.4.3, they are essential tools for the monitoring of landslide phenomena and often, during the performance of forensic investigations, the geomorphologist in charge requests permission to entrust them to the judicial authorities. The hole can also be used to assess in situ the characteristics of the soil involved in the survey, by means of geotechnical tests (for example, penetrometer, scissometric, pressuremeter, dilatometric, etc.), permeability tests (variable or constant load), geomechanical analysis with optical and acoustic camera and geophysical surveys (log sonic and electrical, natural gamma ray, down hole, up hole, cross hole, etc.).

6.5.7 Investigations by Means of Penetration Tests

The principle on which these tests are based consists in the measurement of the resistance that the ground opposes the penetration of a special conical metal tip, mounted on a pole system. Depending on the mode of transmission of loads to the ground and the equipment used, the penetration tests are divided into two main categories: static penetration tests, also called CPT (*Cone Penetration Test*), and dynamic penetration tests, also called DP (*Dynamic Penetration*).

In static penetration tests, the penetration of the tip into the ground takes place at a constant speed, usually 20 mm a second, by means of the pressure generated by a hydraulic piston which acts on the battery of rods. These end with a conical tip and a cylindrical friction sleeve, both to be driven into the soil. The data this test provides are the resistance at the tip and the lateral resistance that acts on the sleeve.

The instrumentation for static tests is installed on heavy vehicles (excavators or wheel loaders), which contrast with their own weight the resistance offered by the ground on penetration, or on lighter vehicles suitably anchored to the ground by means of the propellers.

The static test, which can be performed with a mechanical or electrical penetrometer, is used on fine-grained deposits, with medium-low consistency, and cannot be used on gravel deposits or high consistency. The data obtained, returned in the form of graphs, allow you to directly obtain some physical and mechanical properties of the crossed ground layers and indirectly the expeditious stratigraphy of the subsurface examined.

In dynamic penetration tests the metal conical tip is driven into the ground with impact, through the fall of a hammer from a constant height, usually equal to some decimeters. The measured data in this type of test is the number of strokes (N) needed to achieve a given standard feed of the tip, which is usually established at 30 cm. Depending on the swing weight of the mass, the dynamic penetration tests are divided into four main types, identified in current use by acronyms coined *ad hoc*:

- *Light* (DPL), with a swing weight of the mass $\leq$10 kg.
- *Medium* (DPM), with a swing weight of the mass between 10 and 40 kg.
- *Heavy* (DPH), with swing weight of the mass between 40 and 60 kg.
- *Super heavy* (DPSH), with a swing weight of the mass $\geq$60 kg.

The light test DPL (*Dynamic Probing Light*) is carried out with a highly portable penetrometer, therefore especially suitable in inaccessible and/or wooded areas which are accessible only on foot. The heavy and super heavy penetrometer are instead mounted on mechanical tracked vehicles or with rubber wheels.

Dynamic tests are performed preferably on granular sandy deposits and fine gravel or even on soft rocks. The test becomes meaningless in the presence of coarse gravel, stone blocks and cohesive sediments. The depth of investigation varies according to the weight of the impact mechanism used: light tests can reach a maximum depth of about 8 m from the ground level; medium ones, between 20 and 25 m; while heavy and super heavy tests can even reach depths of 25 m from the PC.

Data acquired through dynamic tests enable a semi-quantitative estimate of the ground consistency. Through the aid of simple correlations, like those for static tests, it is possible to obtain some physical-mechanical characteristics of granular sediments examined and reconstruct, in an indirect way, a first subsurface stratigraphy.

This last aspect makes penetration tests also particularly useful in the geomorphological field, especially in the reconstruction of the geometry of certain accumulation landforms or surface deposits, which are of some importance with regard to morpho-evolutive aspects of a given area and for application problems that affect it. However, given the essentially indirect nature that characterizes these tests overall, it is essential to verify their reliability by comparison with objective data previously acquired *in situ* through surveys of land and direct soil tests.

Case Study

In countries like Italy, morphogenetic processes on slopes that are due to gravity are very common; for example, one can cite the data provided by the IFFI Project (Inventory Landslide Phenomena in Italy) by ISPRA that has recorded, to date, more than 500,000 landslides. In this context, there are many procedures and legal disputes arising as a result of the occurrence of landslides.

In these cases, the geomorphologist must analyze the slopes subjected to gravitational dynamics to detect the presence of landslides, assess the main elements (geometry, type of movement, speed, and volume) and evolution (state of activity).

The writers wish to provide some suggestions regarding the methods and surveying techniques of these important phenomena to unify and standardize the work of the geomorphologist when engaged in forensic investigations involving landslides.

What is meant by landslide?

The term "landslide" has no definite meaning, as it might seem in appearance. Sometimes this term is used to indicate the displaced material, the so-called body or landslide accumulation, interpreted as the volume of rock that has undergone movement and is deposited at the base of the slope (Fig. 6.21).

Fig. 6.21 Body of landslide of large blocks of metamorphic rocks, partially covered by vegetation (Bellino Valley, Piedmont, Italy)

At other times, however, this term indicates the type of movement that has affected the landslide material that subsequently becomes accumulated; in this case, we speak of rockfall, or flow, etc. (Fig. 6.22).

More rarely, a landslide is identified by the detachment niche (Fig. 6.23).

The most widespread and commonly accepted definition of landslide is that proposed by Cruden (1991), who with this word indicates the generic "movement of a mass of rock, earth or debris down a slope". Taking into account the movement type, you can, by referring to the recommendations of WP/WLI (1990,1993) and Cruden and Varnes (1994), distinguish landslides into the following main categories: fall, topple, slide, spread and flow.

This classification has practical implications, because it indirectly gives some guidance on the lithological characteristics of the rock types involved in the movement and the speed of their movement. Other information can also be added to the type of movement, such as the nature of the material and its speed (for example, a flow landslide can be defined as "slow flow of clay"). It should be pointed out that in reality landslides are often characterized by complex types of movement that are not always attributable to a single type. Deep-seated gravitational slope deformations (DGPV), solifluction, soil creep and debris flow (Fig. 6.24) can be also be assigned to the classification cited above.

Fig. 6.22 Landslide phenomena characterized by a roto-translational movement (upstream area) that evolves into earth flows (the valley area) (Alta Badia, Dolomites, South Tyrol, Italy)

Fig. 6.23 Newly formed detachment niche (Lazio, Italy)

Fig. 6.24 Accumulation of debris flow in the final stretch of the valley dell'Uviale of Patresi (Elba Island, Italy)

Table 6.4 Landslide speed scale

Grade	Description	Speed
1	Extremely slow	Up to 16 mm/year
2	Very slow	16 mm/year to 1.6 m/year
3	Slow	1.6 m/year to 13 m/month
4	Moderate	13 m/month to 1.8 m/h
5	Rapid	From 1.8 m/h to 3 m/min
6	Very rapid	3 m/min to 5 m/s
7	Extremely rapid	>5 m/s

Taken from IUGS/WGL (1995)

Speed of movement

Regarding the speed of the landslide, which is a significant physical mass and must be assessed carefully (please refer to Table 6.4), speed is derived from the scale of intensity of the landslide according to the *International Union of Geological Sciences/Work Group on Landslides* (IUGS/WGL 1995), which in turn refers to what is proposed by Cruden and Varnes (1994).

Activity states

The activity status of a landslide, an issue much discussed in science and in practice, is often the cause of disputes in the forensic field. This concept contains information about the time at which a landslide occurs and allows its evolution to be defined in the temporal sense. The evolution may occur by different mechanisms and with different movement speeds. In some cases, the morphological evidence of the movement can be clearly visible during the field survey activities, but in most cases are only detectable by means of technical instrumentation or by studies and comparative analysis (cf. Par. 6.4). In the first hypothesis, the evolution of the landslide must be characterized by a high speed, perceptible to the geomorphologist during land surveys; in the other hypothesis, the motion is characterized, in general, by very low speeds.

In general, landslides are classified according to their state of activity, into three main classes: active, dormant and stabilized, these classes can in turn be divided into further sub-classes as shown in Table 6.5.

The phrase "at the time of survey", reported the first column of Table 6.5, must be interpreted in the sense of "picture" of the situation at the time of field observation, during survey of the terrain.

In the definitions given in the same table, a landslide can be considered active when it has moved within the last seasonal cycle (maximum 1 year), but it cannot be moving during the study (active suspended). Even landslides defined as dormant

Table 6.5 Activity state of landslide events

Active: landslide in motion at the time of the survey or which moved within the last seasonal cycle	Suspended: a phenomenon that has moved within the last seasonal cycle, but that is not moving at the time of observation
	Reactivated: a phenomenon that is active again after a period of inactivity
Dormant: a landslide not in motion at the time of survey, but that can still be reactivated because the root causes of the movement itself remain	The return times of the phenomenon are extremely variable:
	Between 1 and 10 years (frequent activity);
	Between 10 and 100 years (medium-term recurring activities);
	Between 100 and 1,000 years (long-term recurrent activities);
	Over 1,000 years (very long-term recurring task)
Stabilized: landslide phenomenon that cannot be reactivated by its root causes	Naturally stabilized: phenomenon that is no longer influenced by its natural causes
	Artificially stabilized: a phenomenon that has been protected by its natural causes from anthropogenic stabilization
	Relict: a phenomenon that has developed in geomorphological and/or climatic conditions considerably different from present ones

Modified from Amanti et al. (1996)

are not in motion during the field survey, but are able to be reactivated at any time according to return times that are very variable, although are generally more than a year. After a landslide, has been reactivated and has subsequently stopped, the new equilibrium conditions can be similar to those prior to the movement.

Stabilized landslides cannot be reactivated by the same natural causes that caused them, but reactivation after stabilization work that is not carried out properly cannot be ruled out.

Based on what has been stated, a few things to note in professional practice, during a forensic investigation can be listed:

- The activity status of a landslide is a temporary characteristic, variable in time and, moreover, of conventional type. This state in fact refers to a particular time frame and more precisely the date on which you ran the field survey, especially when specified on a geological or geomorphological map. Therefore, when you refer to a geothematic map in the forensic field, it is important to take into account the date of printing or, even better, the period of execution of field surveys, because sometimes material is published much later than the survey period.
- The danger of a landslide is defined as the probability of the occurrence of a landslide phenomenon, in space and time, with a certain intensity (or magnitude). The division of landslides into active, dormant and stabilized, especially if based on qualitative observations, could implicitly lead to the conclusion that active landslides are those with a higher probability of evolution, in terms of time, compared to the other two categories. This is not always true because a landslide classified and/or mapped as dormant may present a greater danger than a nearby landslide classified as active. In fact, as a result of intense and/or prolonged meteoric precipitation, a dormant landslide can be reactivated sooner (in the temporal sense) than an active landslide that has been momentarily suspended, even if it is not very distant from the first, where the weather conditions have been less adverse.
- The most important elements to consider in a landslide are the volume and speed. These elements, in fact, are used to define the intensity or magnitude of the phenomenon. Thus, a dormant landslide of considerable size and depth when reactivated will have an intensity, and therefore constitutes a greater hazard than a landslide of equal speed that is active but superficial (smaller volume). In turn, a dormant landslide with high movement speed will be more dangerous than an active landslide of equal volume characterized by slow movement.
- When judicial investigations concern a landslide is essential to specify in the technical report to be delivered to the judge or the parties to the dispute which method (qualitative or quantitative) has been used in the surveying and analysis of the characteristics of the landslide.

The methodological approach for the study of a landslide is necessarily interdisciplinary because of the large amount of information to be acquired, which includes: the nature of the displaced material and its physical and mechanical characteristics, the type and speed of the movement, the geometric distribution of the landslide, the depth of the sliding surfaces, the groundwater level, the values of pore pressures, the

M. D'Orefice and R. Graciotti

Fig. 6.25 Main morphological elements (detachment niche, steps, slopes, undulations, swelling, subsidence, counterslopes, cracking, shear planes, etc.) found at the Maierato landslide after the event on 15 February 2010 (City of Vibo Valencia, Calabria, Italy) (Source: P. Di Manna)

activity states and, where possible, the age. This data can be obtained by a series of investigations that include: field geomorphological surveys, photointerpretation and analysis of satellite data, mechanical surveys with continuous core drilling, geotechnical testing in situ and in the laboratory, geophysical prospection, inclinometer type instrumental monitoring, surface and satellite topography, geochronological dating, etc.

The geomorphological survey of the landslide area is mainly based on observations directly gathered on the ground relating to the "freshness" of particular morphological elements (detachment niches, carvings, steps, slopes, undulations, swelling, subsidence, counterslopes, cracks, fissures and lateral shear planes, pressure ridges, stagnant water, etc.) (Figs. 6.25 and 6.26), the presence of the reshaping of a body of landslide by other morphological processes, the existence of necessary and sufficient conditions for a reactivation. In this phase, it will also be important to consider any disturbances produced on arboreal vegetation and the existence of damage to man-made structures, etc. Useful information can also result from interviews with the local population, from historical data archives and the comparison with biblio-cartographic and photographic documents dating from different periods.

For investigations of particular importance and when the definition of the state of activity of the landslide is expressly requested, it is essential to carry out instrumental technical surveying which provides, among other things, adequate deep geotechnical inclinometric and surface topographical monitoring, of a significant duration. Only in this way will it be possible to determine with greater precision the morphodynamic evolution of the phenomenon, the depth of the sliding surfaces, the volume of the body displaced and the movement speed. This assertion is valid especially when the movement is extremely slow, and therefore not perceptible to the human eye.

Fig. 6.26 Considerable swelling of the paved road affected by the Maierato landslide during the event on 15 February 2010 (City of Vibo Valencia, Calabria, Italy) (Source: P. Di Manna)

Chapter 7
Environmental Forensics

Gian Paolo Sommaruga

Abstract Nowadays, an ever-increasing number of companies have violated environmental rules. In this case, the environmental forensic expert figure plays a fundamental and decision-making role, using investigative methods and technologies, beforehand applied only in industrial or military courts. Although science and technology are making giant strides in this context, the success of a forensic environmental investigation still depends on the level of know-how of the expert.

Keywords Forensic environmental investigations • GPS • DTM • Groundwater

7.1 Inspections and Historical Research

In recent decades, there has been a considerable increase in rules and legislation aimed at protecting the environment, as we become consciously aware that repairing the damage is much more difficult and expensive than avoiding causing it in the first place, if not at times impossible in order to restore the original condition of the soil and subsoil.

An ever-increasing number of companies that have violated environmental rules and rely on environmental forensic experts to make decisions to resolve conflict relating to environmental damage, whether these are nominated by individual parties and appointed by magistrates.

In this set the environmental forensic expert figure plays a fundamental and decision-making role, using investigative methods and technologies, beforehand applied only in industrial or military courts.

Although science and technology are making giant strides in this context, the success of an inquiry still depends on the starting points, consisting of the initial surveys and historical research.

The key issue is that the initial site visit should take place as soon as possible, immediately after the commission received by the environmental forensic expert,

G.P. Sommaruga (✉)
GEOLAND, Varese, Italy
e-mail: gisomma@tin.it

© Springer International Publishing AG 2017
R.M. Di Maggio, P.M. Barone (eds.), *Geoscientists at Crime Scenes*,
Soil Forensics, DOI 10.1007/978-3-319-58048-7_7

145

even if the parties involved tend to postpone citing various reasons, including the inability to appoint its own technicians or their inability to participate. In this regard, it is noted that the initial inspection is in most cases carried out in order to inspect the sites, or the occurrence, and therefore does not provide for the execution of unique operations, being limited to visual observations, photographs and filming (Fig. 7.1).

The status of the sites can also change very rapidly, such as for example, in the case of landslides, the successive precipitations cause runoff of the muddy material, erasing also important traces of the movements the land undergoes.

Another change to the state of a site is represented by relief operations, which only in recent years have been documented in detail with photos and video footage. The acquisition of all photos and films of the incident that may have been taken by rescuers, by law enforcement or even the media bodies involved is therefore important.

In the case of environmental issues related to historical pollution, such as contamination of soil and groundwater by an industrial plant that operated on the territory for decades, the first inspection can be performed with more notice and planning, in order to also start the real environmental surveys.

As for historical research, it is important not to omit any of the possible useful sources of information, among which are fundamental:

• Building and town planning maps and documentation;
• Climate and weather data in time series;

Fig. 7.1 2010 Massa (I) landslide of dwelling subject to prosecution with deaths

10.47.23 (+1,0 hrs) Dir=NNE Lat=44,04562 Lon=10,12662 Alt=148m MSL WGS 84

Fig. 7.2 2010 Massa (I) photo in landslide slope, subject to prosecution with deaths

- Taking witness information from residents in the area or people informed of the facts (for example, a waiter passing every day on a stretch of road to get to work, may report strange events observed in the days before);
- Aerial photos from public bodies and flight companies who have worked to acquire the basis for aerophotogrammetric interpretations;
- Reporters and journalists, who may have taken important details for investigation, but discarded them as they were considered to be of little relevance to their journalistic purposes;
- Register of deaths in parishes, for the historical secular frequency of landslides and avalanches (Fig. 7.2).

7.2 Aerial Photos and Satellite Footage

The aerial photos are of vital importance in environmental investigations, as forensic environmental experts can draw from them a great deal of useful information, such as the genesis of a landfill in the various construction phases of the individual lots.

Figure 7.3 highlight the area of interest through a series of orthophotos with overlapping of the contour lines every meter. This processing also allows you to

Fig. 7.3 Thirty-year evolution of the territory from natural watershed in landfill

Fig. 7.3 (continued)

calculate the volumetric differences between different periods investigated accurately and objectively. In addition, the digitization of the contour lines thus obtained also allows the construction of a digital terrain model, as we will see later.

In addition to research and acquisition of existing aerial photos from public bodies or special sites (e.g., www.terraitaly.it, www.regione.toscana.it/web/geoblog/foto-aeree, www.igmi.org/voli/), the forensic environmental expert draws on the assistance of the Judicial Police, requiring special flights to view and capture images of the current state of the site, even in areas that are logistically difficult to access (Fig. 7.4).

During the execution of the environmental investigative operations the forensic expert can also search for high spots around the area that is the subject of his activities, such as taking photographs that will be very useful both during the planning and drafting of the expertise both in the hearing stage to better explain the situation to the magistrates (Fig. 7.5).

In recent years, satellite imagery is becoming increasingly important for tests designed to demonstrate the accusatory or defensive hypothesis in the context of environmental litigation.

By overlapping territorial borders and environmental restrictions that affect it, you can highlight and demonstrate the actual occupation of the territory by specific activities, even in the time series (Fig. 7.6).

Fig. 7.4 Storage and handling waste in an uncontrolled area

Fig. 7.5 View of the area from the operator cabin of the mechanical spider

Fig. 7.6 Environmental restrictions in mining area

7.3 GPS Surveys and DTM

Current technology tools available to environmental forensic experts working on land surveys are used to represent and dimension in an extremely precise and also relatively fast manner.

The combination of GPS survey instrumentation is also adopted also in the acquisition phase in the field of geophysical data, such as electromagnetometric measurement for the detection of buried objects and masses in large areas (Fig. 7.7).

The use of GPS tracking technology allows forensic environmental expert to print precise plans of the surveyed areas, with volumetric calculations of the territory also deferred over time.

Furthermore, this methodology allows for quickly and accurately georeferenced surveys that are carried out, the position of excavations and surveys on the location of structures, buildings, materials, waste, supply points samples and more (Fig. 7.8).

In some cases, however, it is not possible to use GPS methodology and traditional optical instrumentation is used, which allows the detection of vertical walls with prism and the pole that would otherwise be inaccessible, through the use of long range laser distance meters (Fig. 7.9).

Whether traditional surveys with optical instrumentation and laser are conducted, or this is done via GPS, the forensic environmental expert has a clear set of digitized data available, with which he may, at any time return to the field for further investi-

Fig. 7.7 Electromagnetometric relief with GPS georeferencing

Fig. 7.8 Relief with GPS georeferencing of excavations and trenches

gation, represent with appropriate plans the operations undertaken so that they are also perfectly clear to the magistrates, and proceed to areal and volumetric calculations by building the digital terrain model DTM (Fig. 7.10).

The digital terrain model (DTM) is obtained by the interpolation of the contour lines and shows the development of the geodesic surface.

Fig. 7.9 Relief in rock walls using laser distance meter

Fig. 7.10 GPS survey with continuous tracking with off-road vehicle

The DTM is obtained by interpolating the topographic data acquired with GPS, after selecting and deleting of the points that do not refer to land dimensions, e.g., buildings, investigative excavations, man-made structures, electrical cabinets, etc.

Digital processing and volumetric calculations are performed with special software that allows, in addition to processing graphics shown in the figure below, precise absolute volumetric calculations relative to past or planned elevations (Fig. 7.11).

The combination of GPS survey technology in digital echo sounders allows precision bathymetric surveys to be performed in a short time, by means of acquisition of a high number of points under continuous monitoring (Figs. 7.12 and 7.13).

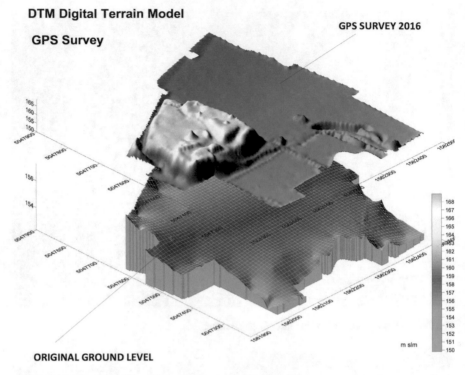

Fig. 7.11 DTM, Digital Terrain Model of uncontrolled landfill

Fig. 7.12 GPS bathymetric relief with continuous monitoring of vessel

By performing GPS survey during the investigation you can develop a DTM of the area under investigation and calculate, for example, the volume of the accumulated waste in sequestration (Figs. 7.14 and 7.15).

DTM - Bathymetrical survey 2016 vs 2015

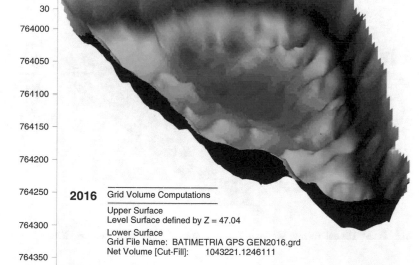

Fig. 7.13 DTM of bathymetric GPS survey

Fig. 7.14 GPS survey of the material under examination

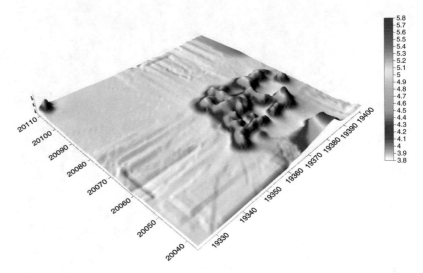

Fig. 7.15 DTM, Digital Terrain Model of the accumulated waste under forensic investigation

7.4   Operating Modes of Sample Collection

The forensic environmental expert must accurately plan both the type of investigation to be carried out and the techniques and investigative equipment.

In the case of investigations related to possible contamination of the subsoil, both in the unsaturated zone and the aquifer, local regulations require the comparison of the concentrations found in the samples with reference tables, above which the terrain (soil or unsaturated – dry subsoil) or the groundwater should be considered contaminated and therefore activate the anticipated procedures for the remediation and restoration of the sites.

After proper informed and/or statistical choice of the points to be investigated, according to UNI EN 10802, excavations and trenches are required to explore the subsurface, installing piezometers for monitoring and sampling in the groundwater zone.

First independently and then necessarily contradictory to the bodies (environmental protection agencies), excavations and environmental surveys with ground sampling are performed to assess the state of contamination with possible investigations even below existing buildings and ongoing and non-interruptible activities (Fig. 7.16).

The evaluation can also be easily performed nowadays in sites that are not yet displaced, by operating with drilling equipment that allows the execution of continuous dry coring by rotary percussion, to cause as little disturbance as possible from a geochemical point of view to the sample that is to be taken and analyzed for environmental characterization (Fig. 7.17).

In some cases, the extraction of fine material from the core can be facilitated with the use of air or water pressure, discarding, however, the last part of the core.

Environmental surveys aimed at the withdrawal of land for the evaluation of possible contamination must include the cataloguing of the extracted material including:

- Lithological description
- Photographic documentation
- Records of visual and olfactory evidence
- Compilation of field reports for the sample extraction chain

The regulations make explicit reference to indicated sampling methods (sites to be reclaimed) and therefore homogenization and sieving in the field, with the withdrawal of only 2 cm under-sieve must be scheduled.

There must be at least three aliquots per sample and in any case a sufficient number both in terms of the different granulometric or analytical determinations to be carried out, and to constitute the counter sample for the necessary bodies and agencies involved, as well as that for eventual reopening that is to be kept sealed (Fig. 7.18).

If the characterization takes place with excavations and trenches, or the investigation involves piles of materials or waste, even in controlled landfills, the environ-

Fig. 7.16 Tilt survey of environmental characterization, which investigates the ground below the building in a non-destructive way and without interfering the construction works

mental forensic expert must solve the problem of sampling an amount that is small enough for analysis, but at the same time one that is large enough to be representative of all the mass from which it comes.

Even in the case of liquid substances, the sample cannot be taken simply by dipping a bottle into a collection tank, but must be executed according to a precise documented plan, which guarantees maximum representativeness, for that portion that will constitute the laboratory sample, of the total mass.

The reference norm used for sampling is UNI EN 10802 ("Liquid waste, granulates, pastes and sludges – manual sampling, sample preparation and analysis of

Fig. 7.17 Particular care must be taken in the extraction of soil from the corer, first removing the drilling crown

Fig. 7.18 Extraction with sieving and sample packaging

eluates") for soil and waste in general, and for so-called aggregates, UNI EN 932 (Test methods to determine the general properties of aggregates – sampling methods).

The first specifies a manual sampling process of liquid, granular, paste and sludge waste, in relation to their different physical condition and position. It also specifies the procedures for the preparation and analysis of eluates.

The second specifies methods for aggregate samples from the suppliers preparation and treatment facilities, including stockpiles. The methods described in the norm are also suitable for obtaining primary samples that can be tested separately. Methods are also indicated for the reduction of samples.

The sampling represents the initial phase of the analytical process that will lead to results whose quality is directly linked to that of the sample. For this reason, the sampling is the most delicate phase of the investigation, conditioning the results of all subsequent steps and have a significant influence on the total uncertainty of the analytical results and consequently the outcome of the environmental investigation.

Listed below are the main definitions relating to sampling, in order to make clearer understanding of the concepts expressed in the text below.

- Sampling: the withdrawal method or to set up a sample. [cf. UNI EN 10802 pt. 3.3]
- Sample: portion of material selected from a larger quantity of material. [cf. UNI EN 14899 pt. 3:14]
- Representative sample: sample in which the characteristics of interest are present with appropriate reliability for the purposes of the test program. [cf. UNI EN 10802 pt.3.5.9]
- Composite sample: sample formed by two or more increments/subsamples combined in appropriate proportions, in discrete or continuous manner (mixed composite sample), from which can be obtained the average value of a desired characteristic. [cf. norm UNI EN 10802 pt. 3.5.2]
- In the field or primary sample: quantity (weight or volume) of material obtained by sampling without any sub-sampling. [cf. norm UNI EN 10802 pt. 3.5.7]
- Laboratory sample: sample or subsample(s) sent to the laboratory or received by the laboratory. When the laboratory sample is further prepared (reduced) by dividing, mixing, grinding or by combination of these operations, the result is the test sample. When no preparation is required, the laboratory sample is the test sample. The laboratory sample is the final sample from the point of view of the sample collection, but the initial sample from the point of view of the laboratory. [cf. UNI EN 10802 pt. 3.5.3]
- Cest sample: sample prepared from the laboratory sample, from which test portions are taken for testing or analysis. [cf. norm UNI EN 10802 pt. 3.5.4]
- Increment: individual portion of collected material in a single operation of a sampling device that is not analyzed/examined as a single entity, but is used for the formation of a composite sample. [cf. norm UNI EN 14899 pt. 3:31]

- Random sampling: sampling of n units made in such a way that each unit has the same probability of being drawn. [cf. UNI EN 10802 pt. 3.3.2]
- Probability sampling: sampling conducted according to the principles of statistical sampling. The essential principle of probability sampling is that each particle or population element has an equal chance of being sampled. [cf. norm UNI EN 10802 pt. 3.3.5]
- Conservation of the sample: any procedure to prevent the characteristics of interest of a sample from changing in such a way that the properties under examination are kept stable from the collection phase until the preparation for analysis. [cf. UNI EN 10802 pt. 3:12]
- Population/lot: the totality of the factors considered. [cf. UNI EN 10802 pt. 3:50]
- Test program: complete sequence of operations, from the first phase in which the sampling objectives are defined, to the last phase, in which the data are analyzed in the face of these objectives. [cf. norm UNI EN 10802 pt. 3:55]
- Project manager: person responsible for the development of the sampling and testing program. [cf. norm UNI EN 14899 pt. 3:11]
- Subsample: quantity (mass or volume) of material obtained through processes for ensuring that the features of interest are randomly distributed into parts of equal or unequal size. [cf. norm UNI EN 10802 pt. 3.72]

A subsample can be: (a) a portion of the sample obtained by selection or division; or (b) an individual unit of subpopulation taken as part of the sample; or (c) the final units of the multi-stage sampling.

The sampling is defined as the extraction process of a substance, of a material, of an environmental matrix, of such volume and composition that the properties measured in the sample, thus defined, represent, within certain reasonable limits, the same properties as the matrix original.

The sampling activities must be carried out by qualified technical personnel in those activities with proven technical expertise and specific training. In the case of scheduled samples, technicians must follow a sampling plan.

The sampling plan provides specific instructions to the sampler and practices and is designed to: (a) Identify and agree on the sampling plan proposed by consultation with the parties concerned on the basis of the relevant legislation; (b) Define what to sample and where; (c) Define the type of substances to be determined according to the reference regulations; (d) Define the sampling strategy (manual or mechanical sampling, random or systematic, timely or medium composite); (e) Define the sampling techniques (depending on the physical state and the positioning); (f) Define any technical precautions to be observed in field operations; (g) Identify parameters to be measured in the field; (h) Record all the precautions that must be implemented to ensure the operations are conducted safely.

For groundwater samples, in order not to change the natural concentration of the substances present in the aquifer body, it is preferable to use a low-flow sampling technique (Fig. 7.19).

For sampling of granular materials, solid waste or sludge it is first necessary to prepare the sampling plan, according to UNI EN 14899 par. 4. The plan provides the

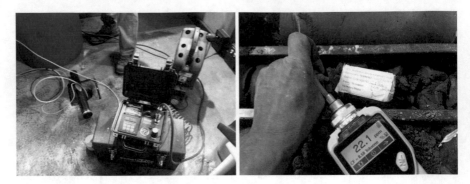

Fig. 7.19 Groundwater withdrawal by low-flow technique and measuring the concentration of Toluene (ppm) with P.I.D. (Photo Ionizator Detector)

sampler with specific unambiguous and practical instructions that, after a visual description of the material to be sampled he records, by means of photographic documentation records, the location and sampling activities.

The employed strategies in procurement may be random, dynamic, systematic or stratified.

In random sampling, the procurement of increments from a batch is performed at random in such a way that each extraction has the same probability of including all test parameters.

Systematic sampling consists in extracting the sample at predetermined spatial and temporal intervals, allowing a uniform distribution of the sampling points.

In the stratigraphic sampling strategy the entire study area is divided into sub-areas (defined as "strata") from each of which a simple random or systematic sampling is taken. This process applies if you want to perform statistical interference of each area.

In the case of homogeneous waste, suitably mixed as a waste liquid within a container equipped with a stirrer, the primary sample is taken from any point within the mass.

In the case of small lots of solid waste, the whole lot is mixed manually or by means of suitable machines and the quartering method is applied to the homogenized batch.

In the case of a single batch of solid waste, proceed with volumetric reduction if necessary by applying the quartering method. If volume reduction is not deemed necessary, proceed to the preparation of a secondary sample through the increment method.

The person in charge of the sampling plan, who if the expert forensic investigations coincide with those of the environmental expert, locates during the methodology planning stage the most suitable method depending on the type of waste to be sampled.

Waste samples must be accompanied by a special sampling and/or forensic report.

The minimum number of increments to be taken in a lot depends on the mass of the lot, the mass of increments and the size of materials which must be taken in addition to the planned physical-chemical analyzes.

In principle, you can follow the following guidelines.

First of all, the mass of increments. The mass of each increment (the dimensions of which must not be less than 1–2 kg for materials with bulk density around 1) is established by the sample extractor depending on the size of the material and the bulk density of the material to be sampled (tons per cubic metre). In the case of manual sampling of particulate material, the minimum mass of increments (I) is calculated by the formula:

$$I = 2,7 \times 10^{-5} rd^3$$

where:

d is the size of the material to be sampled (mm);

r is the apparent density of the "bulk density" material (tons per cubic metre). (cf. par. 4.4.1 of the norm UNI 10802:2004).

The minimum number of increments of bulk material to be collected is determined depending on the volume according to the following table.

Volume (m³)	Increments
<2000	20
2000–3000	25
3000–4000	30

The primary sample is composed of all increments.

For larger volumes of 4000 m³, proceed to the taking of multiple samples considering the whole divided into several homogeneous lots.

In the case of materials or waste arranged on a layer in contact with a large surface area, sampling is performed by means of a grid, with variable sides, depending on the surface occupied by the layer, to form the heap on which operate according to the procedures provided in the previous diagram.

The sampling points of the increments can be provided at the nodes (systematic location) or within the mesh (systematic random location).

Surface m²	Withdrawal points n
<10,000	20
10,000–50,000	21/25
50,000–250,000	26–60
250,000–500,000	60–120
> 500,000	20 per 10,000 m²

In order to obtain the primary sample, the individual increments are thoroughly mixed, so as to obtain a homogeneous mass of its characteristics and a defined composite sample (see. p.to. 3.5.2 norm UNI EN 10802).

Fig. 7.20 Quartering
diagram

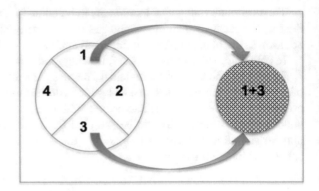

The mixing of solids can be carried out:

– On a towel by placing the material in a pile and rolling over and over with a
  trowel or by mechanical means;
– Inside a bag by applying appropriate movements from the outside so as to mix
  the material.

The mixing of liquids can instead be obtained using appropriate containers and
mixing equipment.

If there is such a large volume of waste as to suffer a reduction in volume, proceed with the quartering method until the necessary volume is reached to obtain the
laboratory sample. (See. p.to. 3.5.3 norm UNI EN 10802).

To perform the quartering, the aggregate sample is placed on the work surface to
be mixed thoroughly, gathering it into a cone shape, then overturning it with a spatula to form a new cone.

Repeat this three times. In forming the cone deposit each trowelful on top of the
new cone in such a way that the aggregate flows around the cone itself and that the
aggregates with different sizes are uniformly distributed and the various grain sizes
mixed well.

Flatten the third cone by inserting the trowel repeatedly and vertically into the
top of the cone, forming a thick flat pile that is uniform in diameter.

Quarter the pile flat along two diagonals, intersecting it at right angles (see figure
below). Discard a pair of opposite quarters and form a pile with the remainder
(Fig. 7.20).

Repeat the process of mixing and quartering to obtain the quantity required for
the laboratory sample (Figs. 7.21, 7.22, and 7.23).

7.5 Packaging and Transport

The methods of packaging and transport of samples are of particular importance in
the environmental characterization of an area, as they must be conducted so as to
ensure the representativeness of the sample material and the reproducibility of the
analysis in other laboratories also at different times.

Fig. 7.21 Deposit some increments onto cloth for the formation of the primary sample

Fig. 7.22 Preparation of the bulk sample to be quartered and initial separation

Usually the soil is sampled in glass jars (possibly new) with a minimum capacity of 0.5–0.6 l (about 1.0 kg) and sealed with a stopper.

To better determine the volatile parts of the substances to be found in soil samples (light hydrocarbons, solvents), it is necessary to withdraw into previously prepared vials that are specifically "calibrated" or containing methanol (Figs. 7.24, 7.25, and 7.26).

In some cases, the material to be sampled is already visibly homogeneous and the removal is done manually by gathering it up in several places with a clean trowel, or directly with hands protected by new disposable gloves, to form the aggregate sample according to UNI EN 932-1, from different points at different depths distributed throughout the lot or the excavation face (Fig. 7.27).

The analysis will be conducted according to the protocol in force at each laboratory; a final ACCREDIA test report will then be drawn up in line with UNI norms.

For each substance to be determined, prior to characterization and possibly also to be indicated in the plan of use, the laboratory must be requested to communicate the amount of material and/or the number of suitable containers necessary for carrying out the analysis.

Fig. 7.23 Quartering on a sample realized in laboratory for the geotechnical (granulometry) and chemical (migration test) analyses

Fig. 7.24 Manual sampling from a meander sediment

Fig. 7.25 Subdivision of the laboratory sample into three homogeneous aliquots

Fig. 7.26 Sampling in a 40 ml vial prepared in the laboratory

The samples will have to be delivered to the laboratory accompanied by "Chain of Custody" documentation of custody, already commonly in use in laboratories, filling out clearly the following information:

"Customer", "Offer", "Project", "Project Coordinator" (the contact person to whom the results are to be sent), "Site/Location" (the sampling site), "Sample Customer ID" (unique description of the sample that will be transcribed on the test report), "Collection Date", "Collection Time", "Matrix" (indicating whether it is a soil, waste, water or air sample), "Number of Containers/Aliquots" (specify number and type of containers for each sample) "Analysis" (for each sample, specify the requested analyzes) (Figs. 7.28 and 7.29).

Fig. 7.27 Manual withdrawal from the excavation face and excavator bucket

7.6   Worksite Operating Issues

While you should give credit to the current regulations for having defined some aspects, such as the definition of technical and operational directions for environmental characterization, on the other hand, the possibility remains of different and conflicting interpretations about the possibility that materials mixed with foreign substances of anthropogenic origin can be excavated just as terrain and be considered by-products (Fig. 7.30).

According to regulations, if the presence of backfill is assumed, during the environmental characterization stage, the following provisions shall apply:

Fig. 7.28 Laboratory
analysis

Fig. 7.29 Safe sample of
unknown dangerous
substance

– Programming the sampling points so as to significantly investigate each portion
  of soil concerned by carry-overs, both vertically and horizontally;
– The percent by mass of man-made elements.

This last assessment, given that in most cases the origin of the materials that
constitute it is not even known a priori, immediately appears problematic and
extremely subjective, since the presence of foreign anthropogenic material that is
over 2 cm in size is difficult to quantify in percentage and by mass if it is not totally
extracted and weighed (together with the weighing of the remaining land) and as it

Fig. 7.30 Excavation area with the presence of carryovers

Fig. 7.31 Backfill with typical presence of bricks

is not sievable (imagine putting a plastic tube of two meters, an entire brick or even milk tainted with paint on a sieve with a 2 cm mesh!). This would require complex and costly operations for complete and accurate quantification.

Also, if the backfill is sieved, as provided for in the technical annexes mentioned by the regulations, any hazardous substances or those that do not conform to legal concentration limits would remain above sieve (physically on the sieve, or because it is not physically taken into the sample that is sieved), resulting in a carryover sample that is seemingly compliant, but that is lacking in its most essential specific parts, those of anthropogenic origin (Fig. 7.31).

7.7 Forensic Environmental Expert's Activities

The forensic environmental expert's activities are exercised on specific assignments, which can be for a public or private body (administration, organisation, individual client or company) or for the judicial authorities. In the former case, they are always defined as a Technical Consulting Party, in the latter, an expert may be required in the areas of civil litigation (torts) or criminal proceedings (criminal investigations).

In civil disputes, the expert is called by a judge to answer the necessary questions on specific matters of fact, and in such cases, assumes the role of T.C. (Technical Consultant), while in criminal matters, they may be called by the Prosecution (Judicial Police, the Public Prosecutor) or Officials (Judges, where the consultant is called a Judicial Expert).

It is precisely in criminal matters that an environmental forensic expert undertakes their core business, playing an important role in the investigations, alongside other professionals like engineers, architects, computer experts, etc.

Moreover, in the context of judicial investigations for criminal proceedings, the forensic environmental expert draws on the assistance of the Judicial Police staff, freelancers and the companies they use, carrying out completely independent investigations (pursuant to Art. 359 of the Codice di Procedura Civile [Civil Procedures Code]) or after discussions with the parties involved, according to the rules governed by Art. 360 of the Civil Procedures Code (Fig. 7.32).

In the block diagram of the following figure, the roles that forensic environmental experts can play in judicial investigations are shown, in both civil and criminal cases (Fig. 7.33).

Fig. 7.32 Forensic geology investigations with the support of the Judicial Police

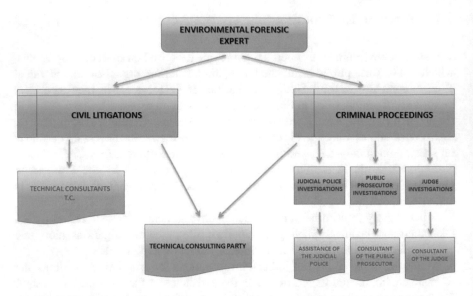

Fig. 7.33 Diagram of environmental forensic expert's roles in a judiciary context

Case Study As a consequence of the discovery of high concentrations of Trichloroethilene (TCE) and polichlorobutadiencs (compounds associated to the production of TCE) in the water of some public wells in the municipality of Limbiate (Milan), detailed researches have been carried out about the causes of the pollution and the possibilities of decontamination. This case illustrates the surveys ordered by the Public Prosecutor of Monza (Milan) related to the ascertainment of the responsibilities.

The contaminated wells lie downgradient in respect to the Cesano Maderno chemical complex, built and used by ACNA Spa (Azienda Coloranti Nazionali ed Affini) at the end of the nineteenth century and at present seat of some chemical plants. During this century, many productive activities (both civilian and military) have taken place in the area. At present, there are many industrial chemical activities which have partially inherited the old infrastructures from ACNA; TCE is no more produced, neither information about the existence of ancient TCE production plants were available at the beginning of the investigations.

The researches have followed this scheme: (i) identification of the potential inclusion centers, i.e., of the plants where TCE could be produced or used, through historical analysis and geophysical (GPR) prospections; (ii) definition of a conceptual model of the local geology and hydrogeology; (iii) investigations of unsaturated subsoil in the risk areas; (iv) drilling of a borehole with full core recovery, into which a piezometer was installed, located in the center of the contaminated area, and chemical analyses of the samples in order to verify the modalities of the percolation of the contaminant to groundwater; (v) hydrochemical surveys finalized to define the contamination distribution; (vi) analysis of the historical series of chemical data referred to all the public and private wells in the area; and (vii) drafting of a contaminant-diffusion map. The research has been carried out using widely the stratigraphical and chemical data provided by local Health Authorities (Gruppo di Lavoro Regionale ex-ACNA).

The main results highlighted the detailed hydrogeological profiles showing the evidence of several low permeability layers within the unsaturated subsoil. Their thickness and extension, however, are not sufficient to create a barrier against the downward migration of the contaminants. Furthermore, some paleovalleys (with reduction of the thickness of the underlying impermeable levels, especially in the area of TCE production) were detected in the impervious base of the phreatic aquifer. This subsoil features could have locally conditioned the course of the contaminants and allowed the migration from the phreatic to the first confined aquifer.

Moreover, the analytical data from soil samples indicated significant presence of TCE practically along all the explored verticals, with a maximum concentration (>3600 ppb) at 20 m below grade. In general, due to the different adsorption capacities, the highest values were detected in the cohesive layers, the lowest in coarse sediments. TCE was detected too, and the ratio between the two compounds is about constant in all the samples. Water samples also indicated remarkable concentration of TCE and related compounds with maximum value for the main contaminant over 3600 ppb.

Finally, the hydrochemical data of the territory surrounding the chemical industry complex were considered. The data highlighted that TCE values upgradient in respect to the relevant area are low (<5 ppb). TCE diffusion begins in the area of the old plants, following initially the geometry of the basis of the phreatic aquifer (as verified in similar cases in Lombardian plain), then the piezometric gradient, and finally arriving to the public and private wells $1.5 \div 4$ km downgradient. Interruptions in the continuity of the protection level of the confined aquifer due to natural (paleovalleys) or artificial (wells) causes determine the contamination of the lower aquifer, which is exploited by several drinking water wells.

In conclusion, the first contamination of a public well (so-called Limbiate 4, about 1400 m downgradient in respect to the contaminated sites) began before 1985 and it reached the maximum in 1994, on the basis of the data from the Health Authorities. The contamination touched progressively other wells as far as 2.5 km further downgradient, but some other contamination sources within the plume are possible.

After 14 years, the Court released the guilty verdict against the private company managing the ACNA area in the 1970s (Figs. 7.34 and 7.35).

Fig. 7.34 ACNA area in the 1937 (*left*) and today (*right*)

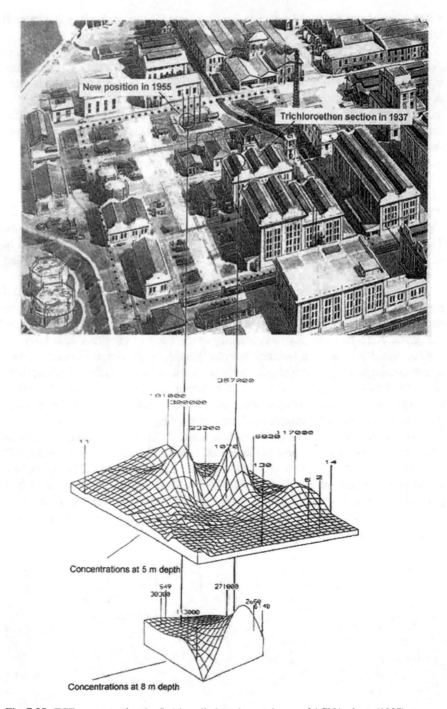

Fig. 7.35 TCE concentration (μg/kg) in soil plotted on a picture of ACNA plants (1937)

Chapter 8
Forensic Geophysics

Pier Matteo Barone

Abstract Forensic geophysics involves the study, search, localisation and mapping of buried objects or elements within soil, buildings, or water using geophysics tools for legal purposes. Various geophysical techniques can be used for forensic investigations in which the targets are buried and have different dimensions. Geophysical methods have the potential to aid in the search and recovery of these targets because they can non-destructively and rapidly examine large areas where a suspect, an illegal burial, landfill or some other forensic target is hidden.

Keywords Forensic geophysics • GPR • NDT • CSI • Accuracy

Regardless of the type of crime and the mode of intervention on a crime scene, recently the approach to the crime scene has evolved with the help of forensic geoscience. Studying, understanding and interpreting the environment and the territory in which the crime itself took place are essential to give a proper direction to the investigative inquiries. This is because the forensic geoscientist must be able to search for and collect adequate information from the environment, placing it in its specific criminal context and characterizing its actual or presumed narrative.

In particular, this approach, in the first instance, provides for the use of various analysis tools in different scale and totally non-invasive in nature. Geophysics, on a local scale, certainly allows the localization and precise mapping of objects (e.g. metal barrels or weapons), bodies (e.g. burials) or cavities (e.g. bunkers), of various types and sizes, concealed underground or underwater (generally forensic geophysics involves the use of electromagnetic instruments, such as ground penetrating radar or metal detector). In concert with these instruments, cadaver dogs are essential to complete the primary investigative framework, providing useful information to get a clearer picture of the geophysical investigation in question.

Geophysical surveys are based on the measurement techniques of classical and modern experimental physics, and in forensic science these can be used primarily for the detection of objects and bodies buried in the ground or pollutants dispersed

P.M. Barone (✉)
Archaeology and Classics Program, American University of Rome, Rome, Italy

Geoscienze Forensi Italia® – Forensic Geoscience Italy, Rome, Italy
e-mail: p.barone@aur.edu

© Springer International Publishing AG 2017   175
R.M. Di Maggio, P.M. Barone (eds.), *Geoscientists at Crime Scenes*,
Soil Forensics, DOI 10.1007/978-3-319-58048-7_8

in the soil, illegal landfills, hidden chambers, armored shelters, etc. The correct use of these methods requires, first of all, terminological clarity to avoid any ambiguity in the definition of investigative technique. In Italian and international literature on the application of physical methods for investigative and judicial investigations, there exist, in fact, some confusion in the use of terms related to remote sensing and geophysical exploration. Broadly speaking, the English term *remote sensing* refers to measurement methods that acquire information, in small or large scales, relative to an object or a phenomenon, using special devices that are not in physical contact with the investigated object itself.

With the advent of new disciplines such as forensicanthropology, that deal with the recovery and study of human remains, there is an increasingly clear perception of the multidisciplinary nature of this field. This multidisciplinary approach is especially evident when it comes to "finding" the victim. As mentioned above, one of the search methods is the cadaver dog. The role of the cadaver dog, until recently nonexistent in Italy, has for decades been a very important human remains search team component in Anglo-Saxon countries and Northern Europe.

8.1   Forensic Geophysics

The physical principle according to which these techniques "see" the buried object is based on the location of the boundaries between the object itself and the surrounding material, and is a direct consequence of the sharp contrast between their physical properties. In practice, from the surface a certain physical quantity of the object is measured, which is a function of a physical property which differs quantitatively from the material in which it is immersed.

Forensic geophysics is the localization and mapping of objects, bodies or cavities, of various types and sizes, hidden underground or under water, using specific geophysics tools, for judicial purposes. In forensic investigations, a wide range of geophysical techniques are applied that have the potential to verify a contrast in physical properties between a target and the material in which it is buried.

Generally, forensic geophysics involves the use of electromagnetic equipment, such as GPR (*Ground Penetrating Radar*) and the metal detector, by means of which it is possible to obtain an approximate image of the variation of physical properties in the first few meters under the surface.

There are many different ways to measure indirect variables related to the physical properties of natural and artificial materials buried underground.

Generally, abnormalities in the variations of the physical parameters can be potentially interpreted as coming from "foreign" buried materials. With these techniques, it is therefore possible to identify and define precisely the place of concealment of the target in question, even to the extent of recognizing evidence of human occupation, anthropogenic disturbance of the soil or soilexcavation, both recent and after many years. The geophysical methods have the potential to swiftly and non-invasively investigate extensive areas where someone has tried to conceal an underground clandestine burial or, in general, a forensictarget.

This apparent redundancy of techniques, capable of identifying a given type of buried target, is a strength in geophysics exploration because, as will be discussed in detail later, the ability to choose and apply different methodologies greatly reduces the risk of failure to achieve targetidentification (false negative). Table 8.1 summarizes the main geophysical techniques used in forensic field along with the corresponding target generating the contrast and thus the anomaly to be detected. The choice of the most appropriate geophysical technique for investigation purposes depends on many factors, the most important of which is obviously represented by the technical ability to identify and locate the geophysical target, which is the physical source of the anomaly produced by the buried object to be identified.

However, the possibility of success in a search also depends on other factors related to the geological and environmental context, namely the physical properties of the soil (or materials) in which the geophysical targets are buried. This is particularly delicate and deserves more in-depth considerations, which we shall return later.

The geophysical methods reported in Table 8.1 can be divided into two main groups:

- Passive measurement methods – that which does not disturb the medium under investigation, such as those carried out to estimate the variations of the natural fields associated with the Earth (for example the gravimetric method);
- Active measuring methods – where the object of investigation is perturbed, by measuring the response of the ground to the generation of artificial signals placed in the subsoil (for example, the electrical method).

Note that perturbation of the system is defined as the process of generation and propagation of signals within the medium under investigation, in order to induce a physical response from the system, that response is analyzed to estimate the physical properties of the system itself. The term perturbation is understood here in the physical sense, and does not, therefore, bear any relation to harmful or destructive actions. The opportunity to use a passive or active method during a search depends, in addition to the above-mentioned intrinsic ability to detect the presence of the searched target, also by the environmental conditions in which the search is carried out. In fact, each measurement technique is subject to noise (natural or artificial), which severely limits the use of the instrumentation in the presence of specific disorders as in the case of electromagnetic methods near the power lines.

Table 8.1 The comparison between the various geophysical methods and their forensictargets

Forensic targets	GPR	Electrical resistivity	Magnetometer	Metal detector
Buried bodies	b	a	a	
Explosives	b			
Buried drugs/money caches	b			
Guns and knives	b		a	a
Artefacts	b	a		a (only metallic)
Tunnels/bunkers	b	a	a	
Landfill	b	b		

[a]Recommended
[b]Recommended and High Resolution

Finally, it is important to emphasize that the information produced by the various techniques are presented in different ways: some methods allow the reconstruction above the area of investigation of a two-dimensional map of the physical abnormalities of the measured parameter (for example, gravity acceleration g), others instead allow for "X-rays" of the subsoil (e.g., GPR).

The use of geophysical techniques in forensics is the prerogative of qualified and experienced personnel. However, the general level of knowledge of the operational principles of geophysical approach, the potential and the limits of such methods, should form the cultural heritage of all those who work in investigative and forensic contexts, law enforcement magistrates, lawyers and technical experts. Given the diversity in the technical and vocational training of the different actors involved in the forensic field, it is not easy to find the right approach (and satisfactory for all) to describe the physical principles underlying the operation of these methods. We have chosen, as well, to encourage a simple approach, direct and useful in the discussion about used geophysical methods. Anyone interested in further information on other techniques, can complete their learning by consulting the bibliography.

Forensic geophysics is not only effective for the search and location of weapons or metal drums, burials and bunkers, elements particularly "visible" and distinguishable from more or less homogeneous context of burial or concealment, but it is also very useful in identifying areas with severe chemical pollution, where the contaminants placed in the ground alter physical properties as a function of their concentration and geometric distribution.

Case Report

This case illustrates the use of GPR analysis on a very popular beach. This investigation is currently underway.

The area under investigation was subjected to the demolition of an illegal building erected on the public beach. The investigative hypothesis found that the remains of this demolition (including very dangerous materials, such as asbestos) were buried beneath the sand (Fig. 8.1).

Single GPR profiles were acquired parallel to the shore, starting from a sandy area and finishing in the investigated area because GPR can easily detect the passage from one dielectric constant to another. Moreover, following the investigative hypothesis, a strong scattering phenomenon due to large buried materials is expected in the radargram.

A well-defined stratigraphic anomaly at a depth of 0.5 m is evident in the natural sandy area. Moreover, the intense scattering produced is compatible with buried material, over 0.5 m depth, of a different nature to the surrounding sandy material (Fig. 8.1).

Then, other GPR analyzes were collected 500 m far from the shoreline, in an vegetated area. Based on the police hypothesis, the area under forensic investigation Suffered from the illegal burial of industrial muds from a water purifier on the shore. The aim of the survey was to find this burial site in a private specific field. Single GPR profiles were acquired inside this area, to find anomalies in the subsoil. Based on one of these radargrams (Fig. 8.2), a more detailed grid was acquired as specified in the paragraph above.

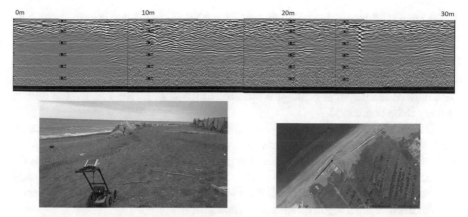

Fig. 8.1 The *bottom* figures illustrated the orientation of the GPR profiles, parallel to the shore-line. The *upper* figure shows one of the radargrams, in which the stratigraphy of the area is clear. In particular, before 10 m along the profile, it is possible to see the natural sandy stratigraphy; after which the radargram detects several anomalies and a strong scattering, starting from 0.5 m deep (v = 0.08 m/ns), confirming the investigative hypothesis

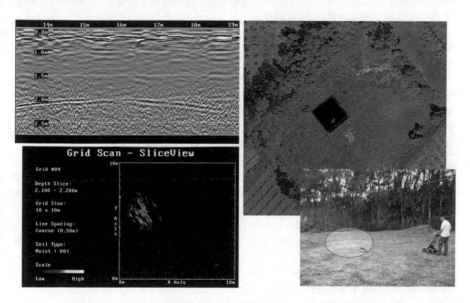

Fig. 8.2 On the *left*, the GPR anomaly is very visible at the same depth (2.15 m) in both the radar-gram (*top left*) and the depth-slice (*bottom left*). On the *right*, the overlap of the GPR map (*top right*) and the location of the anomaly on the surface (*bottom right*)

Both the radargram and the depth-slice clearly show the presence of a well-defined layer at a depth of 2.15 m. Based on its shape and depth, this anomaly could be related to the presence of the illegally dumped muds (Fig. 8.2). Chemical analy-sis of this layer are ongoing to confirm such results.

A similar argument can be made also for human burials, since decomposing bodies release easily identifiable fluids (leachate) that are easily identifiable by geophysical techniques as they alter the physical-chemical properties of the surrounding terrain. Obviously, as with any discipline in the field of forensic geoscience, forensic geophysical investigations must be conducted by experienced personnel, aware not only of the physical principles but also of the notions suitable for the type of survey to be conducted.

The geophysical approach requires, therefore, specialization and experience. In fact, the preparation of the instrumentation, the knowledge and understanding of the subsoil of the area to be investigated, the processing power and interpretation of the collected data are all essential for the success of geophysical campaign. Geophysical methods are very sensitive to small changes in the properties of the ground material, while maintaining a high resolution in the result.

Field data is acquired through a series of measurements performed according to regular profiles and parallel to each other, following a very dense grid determined in advance, related to the size of the target to be detected.

After completing the measurements, the geophysical expert processes information through special software and the resulting analysis of these will produce a map of discovered geophysical anomalies.

The main anomaly that the geophysicist must seek during their investigations is the so-called "cut" that is present in any ground at the moment in which a clandestine excavation is performed.

Such cuts (be they in rock-based or artificial materials such as concrete or cement) persist for centuries from the moment in which they are made. Such cutting is quite evident geophysically thanks to the chemical-physical and compactness differences with respect to the surrounding material, the cut and the filling (whether it is performed with the same type of material or with a different one).

Thus, such an anomaly, is the fault par excellence, always present in any case of clandestine burials, whether related to the burial of bodies, weapons, drums or any other material, and every good geophysicist should be able to detect it in the course o their search activity.

Obviously, any investigation has its peculiarities, so that, at times, together with the cut, in a geophysical analysis it is possible to detect anomalies related to the dispersion in the soil of liquids or metal objects such as shells and cartridges.

To fully understand the operational context is, therefore, of utmost importance when optimizing, not only the methods of choice, but also when focusing on the optimal integration of the techniques mentioned so far.

While the metal detector is limited to the search for metal objects buried at shallow depth, the ground penetrating radar – the best geophysical method in the field of forensic examination – allows for the detection of abnormalities linked to different target types such buried bodies, decomposition fluids, pollutants, bunkers, weapons and any buried object that has a distinct difference with respect to the natural geological background. In addition, GPR has excellent results when used "indoor" or inside homes or huts – built or under construction – to locate concealed entrances, bunkers, crawlspaces or storage spaces hidden under flooring or behind the walls of any material (mortar, reinforced concrete, plasterboard, bricks, etc.) (Fig. 8.3).

Fig. 8.3 The use of the two electromagnetic instruments – ground penetrating radar and metal detectors – at a crime scene

Case Report

"It's about creating a more efficient search"; thus, Jon Dittmar (ERA Technology) commented on a recent investigation that has seen close cooperation between geophysical (GPR specifically) and geo-Forensic Archaeology.

November 2007: in the garden of a house in Irvine Drive, Margate, Kent in England, a lengthy investigation seems to have come to an end. Critical to the research was a ground penetrating radar system that had discovered the bodies of two teenagers, Vicky Hamilton and Dinah McNicol, disappeared since 1991. This geo-physical system, in fact, helped forensicinvestigators and the police to locate the bodies of the two girls.

Jon Dittmar is an engineer of ERA Technology, the company that for the first time in 1994 helped forensicinvestigators and the police in the search for bodies in the home of Fred and Rose West Cromwell Road, Gloucester. He knows not only the power and flexibility of the GPR, but above the absolute decisive effectiveness of close cooperation between the geo-physical approach and other forensic geosciences.

But how can a geo-physical expert understand, analyze and interpret an anomaly as a possible clandestine burial? In short, the GPR, in real time, obtains an electromagnetic response like the one shown in Fig. 8.4. Obviously, each case and each terrain has its own characteristics, but generally, the presence of a hyperbolic event, a horizontal reflector and another hyperbolic event is indicative of digging activity, a "cut" in the ground later, filled, an indelible mark that lasts over time, even for centuries (as archaeology has proven).

Case Report

The scientific investigations at the crime scene have brought to the fore the advantages of geophysics (with particular attention to GPR) for locating and identifying missing persons buried in the ground and not only. Recently, in fact, this method has been tested to see what resolution can be applied to an object, no matter how small, buried underground.

Figure 8.5 shows the results of a scientific investigation in search for a gun concealed at shallow depth in the ground, and a plastic container with drugs and money.

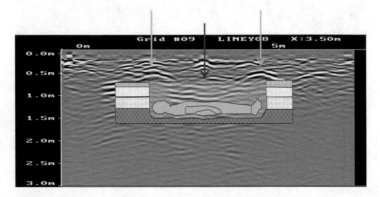

Fig. 8.4 The radargram shows how the *yellow arrows* indicate hyperbole related to the cut made in the soil to obtain the burial; while the *red arrow* indicates the horizontal reflector – filling next to the cut and what remains of the body, with chemical-physical characteristics different from the surrounding soil

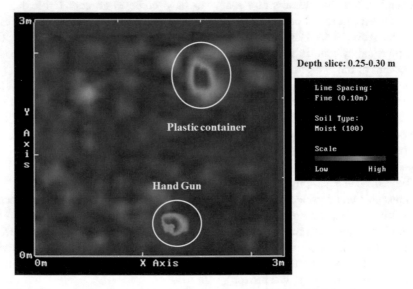

Fig. 8.5 GPR map at the depth of about 30 cm with a clear view highlighting the presence of a gun, and a plastic container

Note how, even though the gun was buried alone, was of small size and wrapped in a plastic bag, the final ground penetrating radar response map allows not only the clear location of the object, but also its geometry. Regarding the container, given its size and the partial presence of air inside (an element that stands out with a very clear electromagnetic response), its location and its dimensions are very clear.

8.2 Pros and Cons of Forensic Geophysics

As mentioned, various techniques can detect the same type of target, although this is not a real equivalence between methods, since each technique "sees" the target in a different way, both because of the physical principle underlying the measurement, and because of the geological and environmental conditions of the area under investigation. The information that results is, therefore, different in terms of dimensions and shapes of the anomalies associated with a same target. The choice of using one technique over another depends on many factors, the most important of which, as repeatedly pointed out, is represented by the ability of the technique to identify and locate the geophysical target in the specific environmental context of its location. This implies that, for a given target, it is not to be taken for granted that a method that has proved successful in a specific context must necessarily produce positive results under different geological and environmental conditions. Thus, for example, the GPR that is the elective technique for searching for objects or bodies inside the masonry and cement, does not provide good results in an electrically conductive material (saturated clays and sediments). Therefore, before proceeding on any investigative hypothesis, it is essential to carry out a careful analysis of the conditions, the nature of the soil and/or the materials that are the object of investigation.

A second factor that affects the choice of method is logistical: the majority of existing geophysical methods can be used only in the open (except GPR). In addition, the particularly complicated topography, the impervious soil structure or the nature of the rocks and soils, can create insurmountable obstacles to setting the electrodes, to the continuous movement of the metal detector sensors or the low-frequency radar antennas. It follows that a particular technique, for all its reliability from the point of view of its ability to identify a target, may in fact be unusable in a given site for purely practical reasons.

The third factor is the timing of acquisition and data processing. In the case of investigative surveys that take place under severe time restrictions and very pressing work conditions, the only really usable method is the ground penetrating radar, because it displays real-time data. However, for sites related to cold cases, it is possible (and desirable) to use various methods in order to integrate the results and minimize interpretive errors related to the instrumental and methodological limitations of each technique. The multi-technique approach is very effective also in environmental investigations, where one can carry out a first large-scale survey with quick inspection techniques, also mounted on aircraft, on the basis of which it is possible to identify areas of interest to be investigated with a detailed investigation.

Based on the above, the choice of the most suitable geophysical method to solve the specific investigation problem must follow a logical scheme that begins with a hypothesis on the nature of the target, considering all the possible geophysical methods suitable for its detection, and gradually discarding those that do not meet the specific survey requirements related to the nature of the material, environmental and logistical conditions and time constraints.

However, we must still address the most important issue for investigation: what is the nature of the object that produces the detected anomaly? No physical or geophysical technique can provide absolutely certain answers. The impossibility of exactly identifying the nature of a located object is an inherent limit of the type of measurement and not the adopted instrument itself or its incorporated technology. In fact, the anomalies measured with different techniques are not unique: different combinations of shape, depth and contrast can generate very similar abnormalities. Under these conditions the only way to resolve the ambiguity is by the use of additional non-destructive investigative measures such as cadaver dogs or archaeological excavation, which is the ultimate act of verification of the information provided by the geophysical methods.

The role of geophysical techniques in forensic field is, therefore, to identify the areas of greatest interest, thus allowing to spatially limit the extension of the area under investigation and to drastically reduce the working times. Once detected and localized, the anomalies should be verified with more invasive techniques, thus following an approach similar to that used in the exploration for hydrocarbons, where the geophysical technique identifies the possible oil deposit, which is then verified by mechanical sampling.

8.3   Cadaver Dogs

Cadaver dogs (or as referenced by the International term K9)[1] have the specific task of locating the remains of a missing person, and can be a great help non-invasive help for the interpretation of geophysical results before resorting to possibly destructive archaeological digging. Despite extensive research in the technological field, we are unable to find a worthy replacement of the dog that has been used since 1974 in these operations. This was when the first dog was used by the New York Police Department to search for corpses. Having verified that dogs could be great assistants, also the FBI began training dogs for this kind of work, including some *National Search and Rescue Association* volunteers.

The use of cadaver dogs is based on the highly sensitive sense of smell of the animals: in a dog of medium size the total area of the olfactory mucosa is150 cm^2, compared with only 5 cm^2 of an adult man, and it is a tenth of a millimeter thick

---

[1] A term associated with the assonance between the Anglo-Saxon pronunciation of the letter 'K' and the number '9' and the English pronunciation of the word 'dog' – 'canine'

against the human 5 thousandths of a millimeter. Canine olfactory cells are about 225 million in number, compared with 15 million a human being; as a result, the dog has a 500/700 times higher ability to perceive odors compared to humans. Dogs may in fact perceive the odor of 2 mg blood serum or 5 mg of urine, they can sense acetic acid in aqueous solution in a ratio of 1 part in 1 million.

In practice, these animals, with their highly-developed sense of smell and a specific training program, can recognize the typical chemical substances (about a hundred) released by a corpse.

However, for the odor to be detected and therefore signaled, it must exceed a certain threshold and it must remain in the dog's attention span for a sufficient period of time: in other words, it is necessary that the odor exceeds the threshold of stimulation and has a high attention time, or it must be intense enough to trigger a recognition, and sustained over a sufficient period of time for the dog to pay attention to it and signal it. On these functions is based the training of dogs, which involves the use of "bait" of odorous material: such material may be made from pork, blood moistened cloth, human teeth, or chemical substitutes such as cadaverine and putrescine.

The search for a dead body is played out for the dog as a hunting exercise where the dog behaves as team player and the handler is the pack leader; the discovery of the corpse is therefore interpreted as the capture of the prey, and the related sharing of the spoils is replaced by the enjoyment of a preferred reward game given by the handler. It is clear that such activity requires a strong relationship built up over the years between man and animal; already when puppies, the handler can start a first training by searching for hidden toys or getting the dog used to search for items. The basic training is based on searching for buried pork meat, by making of small holes in the ground of about 3 cm in diameter to a depth of about ten meters, a distance of about 30 cm from one another, in a straight line for 10 m and across multiple rows approximately 15 cm of relative misalignment.

The exchanged signal code also requires a special type of training: most dogs are trained to signal a find by stopping and sitting down in front of the odor source and waiting for a reward; such behavior is known as waiting for the "rabbit jump", or the prey's attempt to flee from the located den. The handler, without being seen by the dog, causes his toy to emerge at the point indicated by the dog and thus establishes the reward. Alternatively, the dog may scrape the ground surface, adopting a behavior focused on discovering if the smell increases with the removal of the soil. With other more communicative animals, a behaviour pattern may be established whereby the dog signals the find by moving repeatedly between its location and the handlers position, thus inviting the handler to come and inspect the find. In some cases, the search may be enhanced by pushing T-bars[2] into the ground in order to allow greater odor leakage and its stronger identification.

---

[2] T-bars are T-shaped rods with a length of about 100 cm, diameter of 2.5 cm, hollow inside with 40 cm opening on the side, also used in archaeology to verify the consistency of the soil in depth for the search for anomalies.

This shows how the dog alone cannot solve all the problems: strong collaboration between the investigation teams and dedicated handler training are required.

In conducting a search, for example, it must be taken into account that the dog's concentration span in a targeted exercise is reduced to 10–15 min. Therefore, the search area should be narrowed down to the most probable area where the dog may succeed in locating a smell source. On open ground, an area is divided into corridors of approximately 500 m², laid out perpendicularly to the wind direction. Each of these corridors is considered a sector and is classified with numbers or letters.

In some very rare and special cases the cadaver dog is so highly trained and has reached such a level of sophistication that he can perceive the difference between human and animal remains. Thanks to the conditioning techniques incorporated in the dogs' training, they are able to recognize any chemical be fixed in their olfactory memory by trainer (a case example are the dogs trained by the Canine Unit of the State Police at Milano Malpensa).

Finally, it should be emphasized that these dogs are able to sniff the presence of a corpse even if it is submerged in water. In this case, the dogs do not dive, it is sufficient that they sniff the surface of the water to be sufficiently close to the source. From the boat, they are already able to sense the volatile particles released by a decomposing human body. They may remain in continuous stand-by mode, even dozing, but the moment they enter the cone of the smell and sense the possibility of identifying the chemicals that they are trained to recognize, they immediately give the signal that they have sensed something.

In most occasions, the dogs are the most effective and efficient resource. In combination with other technical resources, the dog can give an exact indication of where to start the excavations, thus preserving the integrity of the finds recovered.

Case Report

In Italy and abroad, cadaver dogs are increasingly finding a role alongside GPR in forensic investigations. Famous are the international case of Madeleine McCann and the Italian case of Marina Arduini. But this concerted action could be increased if the experts involved in these surveys had a preparation of excellence and undisputed.

Few years ago the Forensic Geosciences Italy team was involved in the reopening of a *cold case* located in Marcianise (CE): the disappearance of Pasqualino Porfidia, a 8 years old child. This case caused quite a stir in its time (1990) also involving well-known television programs such as "Chi l'ha visto [Missing]". The reopening of the case, based on new evidence, required the joint use of ground penetrating radar and cadaver dogs at an early stage for the non-destructive and accurate search for anomalies related to burials or concealment of a corpse, in order to later proceed to a more detailed search by means of archaeological excavation. This integrated search identified a specific area in which both the GPR and the cadaver dog, independently gave alert responses. The subsequent archaeological dig provided only a partial confirmation (Fig. 8.6).

Fig. 8.6 A photo of joint
GPR/K9 investigative
operations

8.4   Survey Criticalities

The geophysical approach requires, therefore, specialization and experience. In fact, the preparation of the instrumentation, the knowledge and understanding of the subsoil of the area to be investigated, the skill in processing and interpreting the collected data are all essential for the success of a geophysical campaign, especially in the forensic context. If GPR data acquisition might seem trivial (in fact it is rather complex and delicate), the interpretation of the data itself requires years of experience and countless studies.

Those who order the use of GPR in geophysical surveys (the courts, the police, the prosecutor or the attorney) must be aware of not only its potential, but also its limitations. This is not a magic solution, but a scientific instrument and as such has known measurement uncertainties that must be taken into account at the time of acquisition and interpretation of data.

Likewise, cadaver dogs, while being valuable resources, are living beings and as such, sometimes prone to error if employed in the wrong way. In addition, their full potential has yet to be explored: however, we must remember that the success of the search operations also depends on the organization of the activity, which must take account fatigue and distractions of the dog. Of fundamental importance is therefore the coordination between different dogs and planning activity involving the other figures present, such as the geophysicist, the geologist, the forensic archaeologist or the coroner.

Appendix: How Does GPR Work

In principle, *Ground penetrating radar* consists of sending into the ground high-frequency electromagnetic pulses (10–3000 MHz) and measuring the time taken for the signal emitted from the transmitting antenna to return to the receiver after reflection and/or diffraction from any discontinuities present in the material under investigation. The round-trip time (TWT), expressed in nanoseconds – ns, allows for the measurement of the distance in time between the antennas and the "target"; this distance may be transformed into depth (meters) in the subsoil once the speed of propagation of the pulses in the medium under investigation is known.

The attenuation of these pulses in the subsoil is related to two factors: the presence of moisture in the soil and the chosen frequency. As regards the presence of moisture, a high water level in the soil makes it very conductive and this significantly attenuate the electromagneticsignal with the risk that it does not penetrate (or only partially penetrates).

The choice of frequency to be used depends on whether the transmitter is connected to an antenna (Tx) which produces a very short electromagnetic pulse (on the order of 1–10 ns). The chosen pulse duration, in turn, is linked to the antenna frequency and the required vertical resolution, or the ability to distinguish between two layers or objects close to each other. In other words, the higher the frequency of the antenna, the shorter the pulse, which results in a low signal penetration (because the attenuation also depends on the frequency) but in a higher vertical resolution.

GPR instrumentation typically has two possible configurations: the so-called bistatic configuration, in which the transmitting antenna is physically separate from the receiver; and the monostatic configuration in which the transmitting and receiving antenna coincide.

The graphical representation of GPR data is a fundamental step in the understanding and interpretation of the results. These results are presented as greyscale radargrams (or stratigraphies) of the sub-soil and modern software allows for very high visual resolution and definition. Furthermore, if the ground scans envisaged parallel profiles within a grid, then maps (or plans) may be obtained and displayed of the investigated region that represent, at various depths, not only the geometries of the buried objects but also their size, normally using a medium envelope algorithm, also known as an *average amplitude envelope* (Fig. 8.7).

To correctly interpret a radargram, you must know how the section was acquired. The transmitted pulse from the radar antenna is not propagated in the soil or in a linear manner material like a laser, but rather it behaves as a so-called radiation cone

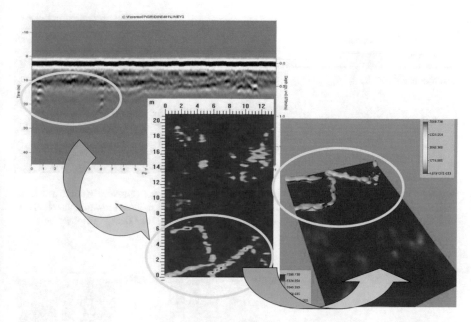

Fig. 8.7 These are examples of how the same anomaly (in this case a wall structure) can be displayed in three different ways: a radargram or em subsoil stratigraphy (*top, left*), a map (or plan) at different depths in the subsurface (at the *center*) and a 3D reconstruction (*bottom, right*) acquired with the GPR

"illuminating" the buried target also before being perpendicularly over the target itself (like a lamp burning in the darkness of a room). The diameter of the cone increases with the depth of the ground penetrating radarsignal. Moreover, its dimensions also depend on the acquisition surface conditions of the antennas frequency used (for example, high frequencies constrict the diameter of the cone) (Fig. 8.8).

The presence in the subsoil of a void or of any more or less point-wise object produces a characteristic electromagnetic response: the diffraction hyperbola. The hyperbolic anomaly comes from the reflection of the point-source (buried target) and occurs, as we have seen, because the energy is emitted in the form of a cone that 'illuminates' an area larger than the target itself. Consequently, the signal is reflected not only in a perpendicular direction to the target directly below the antennas, but also just before and just after, thanks to the additional transmission of oblique waves. Only the hyperbola peak corresponds to the actual position of the source (Fig. 8.9).

The maximum horizontal resolution approximately corresponds to the footprint of the radiation cone (or illuminated area). The signal round trip time, and consequently of the depth estimation, may be calculated using the so-called calibration of the hyperbolic traces resulting from an abnormality. It is important to note, however, that it is possible to determine the depth of a target only if the speed of signal penetration through the material or materials is known.

Fig. 8.8 The transmitting antenna (Tx) emits a signal that travels not vertically in depth as a laser, but it creates a radiation cone that 'illuminates' the target and is reflected towards the receiving antenna (Rx). To the *right*, the analogy with an illumination cone, for example purposes only

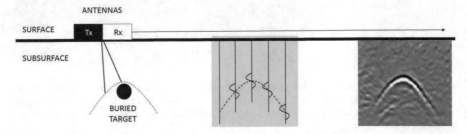

Fig. 8.9 The diffraction hyperbola, visible in this figure, derives from reflection of the intercepted target. It should be noted that only the apex of the hyperbola identifies the correct position of the buried object

With the exception of buried conductive materials (for example, metal, which has a high conductivity and magnetic permittivity), electromagnetic waves pass through the buried target, continuing their penetration and producing different reflections at different depths. In some cases, this effect allows us to estimate not only the depth of the upper part (top) of the object, but also its vertical dimensions (for example, in the presence of an underground tunnel, it is possible to identify not only the top of the tunnel, but also its bottom).

Chapter 9
Forensic Archaeology

Pier Matteo Barone

Abstract Forensic archaeology is a scientific discipline that uses archaeological theory and methodology in a legal context. It combines archaeological, taphonomic and criminalistic knowledge to localise, document and interpret archaeo-pedological, archaeo-ecological and osteological finds and patterns at a (possible) place of incidence or a crime scene. Less famous, but equally important, is the employment of the forensic archeologist use to prove the origin of archaeological finds and works of art subjected to clandestine excavations and illegal trafficking, to draw up expert opinions and assessments of damage to cultural heritage and archaeological finds and confiscated artworks.

Keywords Forensic archaeology • Stratigraphy • Archaeological crimes • Relative chronology • Clandestine excavation

The scope of the forensic archaeologist is varied. The most famous and even more in demand service pertaining to archaeological skills and competence in the legal-medical context is the use of reading and interpretation techniques related to material traces, in particular in the analysis of the crime scene, for the recognition and classification of findings, the identification of their origin and age, and for the reconstruction of the spatial arrangement of people or objects in a given place and time, as well as the time sequence of human and natural activities that took place on the scene of a crime.

Less famous, but equally important, is the employment of the forensic archeologist use to prove the origin of archaeological finds and works of art subjected to clandestine excavations and illegal trafficking, to draw up expert opinions and assessments of damage to cultural heritage and archaeological finds and confiscated artworks. With these terms are also indicated those professional activities related to judicial or extrajudicial matters; for the protection of cultural heritage, archaeologists are engaged as relevant experts or consultants by the judiciary, prosecutors, police forces, institutions, local or private organizations, lawyers and the other

P.M. Barone (✉)
Archaeology and Classics Program, American University of Rome, Rome, Italy

Geoscienze Forensi Italia® – Forensic Geoscience Italy, Rome, Italy
e-mail: p.barone@aur.edu

© Springer International Publishing AG 2017   191
R.M. Di Maggio, P.M. Barone (eds.), *Geoscientists at Crime Scenes*,
Soil Forensics, DOI 10.1007/978-3-319-58048-7_9

figures. Sometimes, the Ministry of Cultural Heritage places legal restrictions on areas with high archaeological potential. Based on this, nobody, including the land-owners, can remove soil for any purpose, from building new construction to plant-ing. In this case, the best option for private landowners is to evaluate and understand the archaeological potential of the area. Forensic archaeologists can help in this and they can work for both sides, landowners and Ministry, to evaluate better the situa-tion of the archaeological restrictions, to completely remove, or to reduce the legal restrictions to better-defined areas.

Archaeology is applicable in the forensic field when there is not only the need to search and locate clandestine graves and recover the remains, but also on the occa-sion of evidence concealment, pollutants and any other underground material. The sequence of operations leading to the location and recovery of buried remains is divided into three main stages. A preliminary survey of the alleged place of burial to determine the location of exactly where do the investigation, the actual excavation phase and the documentation and recovery of the remains as these are excavated.

The archaeologist's duties consist of performing a dig, carefully documenting each step, and recovering discovered objects in the most appropriate manner, whether they consist of human remains or other materials. In a forensic context, it may be important to know how to properly remove any human remains, even if, often, in excavation sites it is the finds consist merely of scattered bones or teeth that have been picked up and moved by predator animals and scavengers. The recovery of remains by forensic archaeology techniques can provide information about the mode of deposition-concealment of corpses and provides clues to the investigators.

The archaeological excavation is a destructive operation; consequently, the need to fully document every single step has always represented a very important factor. Thus, over the years we have seen the development and improvement of more and more accurate methodologies whose main purpose is to maximize the amount of information obtainable from the investigation site. This approach is particularly appropriate and useful in the forensic context, where the investigation site is the scene of a crime and the information stored in it can be of crucial importance for the reconstruction of the events and the success of the investigation. Good excavation documentation may be drawn up by properly trained police officers; however, the analysis of the excavation, of the remains and the interpretation of data and the coordination of subsequent recovery operations remain the exclusive prerogative of forensic archeology specialists.

9.1 Forensic Archeology

Archaeology is the study of the ancient and recent human past through material remains. From the fossil remains of millions of years ago, belonging to our first human ancestors in Africa up to today's twenty-first century buildings, archaeology analyzes the physical remains of the past in search of a broad and comprehensive understanding of human culture.

Forensic archaeology is the application of principles and archaeological methods to locate and recover evidence within the boundaries of the criminal justice system or for use in civil cases, by operating as a discipline in its own right.

Persons qualified to conduct forensic archaeological investigations are trained in traditional archaeological techniques, but they are quite flexible in their approach, adapting these methods to different forensic settings that may arise in criminal or civil proceedings. Integrated skills and considerations beyond the traditional archaeology parameters are: possession of a basic knowledge of law enforcement and legal procedures (as well as the ability to cooperate productively with law enforcement personnel), the ability to conduct efficiently and effectively the investigation, under tight time constraints and under the scrutiny of the media, as well as the ability to navigate skillfully in situations that deviate from the traditional archaeological experience, as for example in the presence of a burial that includes conservation the remains of soft tissue.

The sequence of operations leading to the location and recovery of buried remains is divided into three main stages. A preliminary survey of the suspected burial site in order to determine the location of exactly where to carry out the investigation, the actual excavation phase, and the documentation and recovery of the remains as these are excavated.

Reconnaissance, contextualization and localization involve the interpretation of the terrain in order to search for a presumed burial site and are carried out through the use of different instruments, at both large and small scale levels. For example, aerial photographs can easily provide important information about the area to be investigated (abnormal changes in vegetation cover over large areas or changes in soil morphology) and their use in the preliminary stages of a survey can be effective, leading to significant savings of time and energy. On a local scale, geophysical techniques that allow a more accurate positioning of the buried target (Fig. 9.1).

The archaeologist's duties consist of performing a dig, carefully documenting each step, and recovering discovered objects in the most appropriate manner, whether they consist of human remains or other materials. In a forensic context, it may be important to know how to properly remove any human remains, even if, often, in the excavation sites is very easy to find just scattered bones or teeth that have been picked up and moved by animal scavengers.

The recovery of remains by forensic archaeology techniques can provide information about the mode of deposition-concealment of corpses and provides clues to the investigators. The analysis of the context, and the excavation of the remains, the interpretation of environmental data and coordination of the subsequent recovery operations remain the exclusive prerogative of archaeology specialists. Archaeology is a destructive process and for this reason it is preferable to integrate it with other methods to switch from non-destructive analysis to more invasive ones, in order to minimize the loss of information.

To this end, three basic objectives in an archaeological forensic investigation can be summarized. The first is the understanding and interpretation of taphonomic events - the story of a site after it has been created through the deposition of remains. Taphonomic events are the processes both natural (N-transformations) and cultural

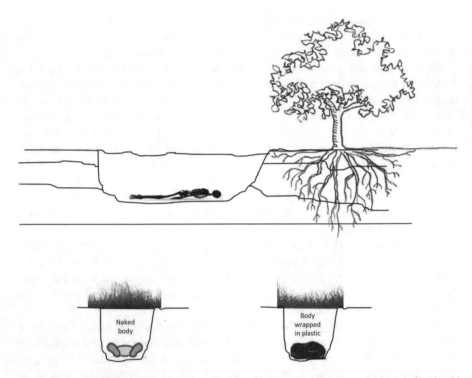

Fig. 9.1 In the figure *above*, we see an example of how the section plan of a burial site should appear; in the figure *below*, you notice how the vegetation changes depending on what is buried in the ground

(C-transformations) that occur in a site and alter or transform it over time. These processes include: N-transformations, such as the seasonal outflow of water, high levels of animal or insect activity, activity and growth of tree roots; and C-transformations, as an informal digging (due to lack of awareness that the location is a place of criminal activity or the action of a civilian), the deposit of waste or other foreign objects in the event of deposition, and high levels human traffic that can disrupt the original deposition contexts. Understanding and being able to interpret the transformations that these processes result in a site are essential elements for proper archaeological investigations.

The second main objective is a reconstruction of the causal and contextual events related to the creation of the site and the deposition of the corpse. This is accomplished through the collection of disturbed surface elements, excavation methods and detailed and photographic documentation covering all phases of the investigation. It must be remembered that the forensic investigation of an archaeological site is inherently destructive and forensic archaeologists must therefore take all precautions to preserve the maximum amount of evidence and context information possible.

Finally, based on the evidence provided by the material record for the narrative reconstruction and the taphonomic history of the site, an investigator is able to achieve the third objective, that of a conscious interpretation of the events that concerning the deposition of the deceased, which will aid in the resolution of the case.

Thus, it is evident that the archaeologists specialized in forensic investigations provide the police with a complete reconstruction of the crime scene with the possibility of investigating both cases which are often very distant in time, as well crime scenes that are still "warm" and very recent, while maintaining the same high standards of evidence recovery and accuracy.

To work in the forensic field, the basic archaeological skills extend well beyond the academic. These include not only know perfect knowledge of excavation techniques, but also competence in pre-excavation (mapping, analysis of soil and remote sensing) and post-excavation investigations. The most important factor, then, is that the archaeologist must have a broad experience background. Archaeologists who have always worked on a single type of site and/or on a single geographic area may have developed specific expertise, but they may yet be unprepared for the variety of situations normally encountered in forensic work. Archaeologists need to be ready to carry out excavations in wells, open fields, woodland, rock shelters, caves, in burned areas and not, latrines, waste landfills, urban contexts (such as underground utilities, infrastructure, buildings, etc.) and cemeteries.

These challenges highlight a fundamental issue, namely the need for flexibility and experience in a very broad range of site types. Forensic work is not the occasion to be initiated into the exhumation of skeletal remains or the analysis of illegal dumping sites. The work proceeds at a pace that is too fast for learning purposes, and the result is too important to risk that beginners make mistakes on site.

Also, strong preparation in osteology is useful for the archaeologist. For example, in sites with several diversified skeletal remains, it is normal to separate them based on the spatial arrangement, age and sex of the individual. If the remains are not well catalogued as they are removed from the ground quickly and without due care, the chances of error in the post-mortem examination become much larger.

Finally, archaeologists who engage in forensic work need a thorough understanding of the types of artifacts that may be encountered. In forensic work, this means that they need a solid grounding in modern weapons that are common to the area as well as a basic knowledge of explosives and ammunition. In the exhumation of war dead, this knowledge requirement also extends to grenades, anti-personnel mines and other explosive devices, which require identification and documentation. This particular knowledge serves a dual purpose: to identify and circumscribe unexploded ordnance, and ensuring that it does not explode, for the preservation of personal safety.

That is needed, then, is also a vast knowledge of modern materials culture, as well as their use and history. All kinds of modern objects must be recognized from the fragments typically found in forensic sites. This includes not only the archaeologist's own cultural background, but also that of all the other ethnic groups that might be present in the investigated site. For example, in the most popular cultures (for most of us) it is common, for married people, to wear a ring on the fourth finger of the left hand: these rings often bear inscriptions with dates and initials of the names of the persons involved.

If this is a very simple example, in any case we must emphasize the need to become familiar with the material culture of other ethnic groups. For example, a small package containing a folded piece of paper can be an amulet with a Muslim prayer, a clue that could help shed light on ethnicity, religion, and, perhaps, the identification of the wearer.

Furthermore, also a wide knowledge of the conservation techniques of finds as exhibits is necessary for forensic work. Much of the material found in forensic cases will be extremely perishable material that is usually not normal for an archaeological context, such as clothing, personal documents, paper money and religious or ethnic objects. Such material is likely to suffer rapid deterioration and may need immediate and extensive action to stabilize the degradation, conservation and safe-keeping measures for removal and further transport in analysis laboratories.

Of course, the archaeologist needs to work in these contexts, collaborating with other experts of the scientific investigations, ensuring the professionalism, care and caution necessary to obtain evidence without compromising its validity.

Archaeological case history is, therefore, as for technique, extremely useful in investigative work, with particular regard to the context. In fact, there is a lot more information in the context in which an object was found rather than just in the object itself. The context may involve other artifacts, local peculiarities and natural environments (Fig. 9.2).

Fig. 9.2 Some archaeological excavation phases during a forensic investigation

9.2   Stratigraphic Excavations at the Crime Scene

The success of an archaeological dig begins long before the discovery of the body or site to be investigated. A well-defined protocol must always be followed, staff must be trained with the help of experts and appropriate equipment must be sourced. Each case has its own peculiarities, but usually to meet the first two requirements, it would be appropriate to follow in principle the scheme illustrated in the following table (Table 9.1).

Moving onto the actual excavation stages, during the investigation it is assumed that the forensic archaeologist has full knowledge of the soil composition present at the crime scene to assess at first sight what is abnormal. Soil formation can be caused by several material factors, climatic, topographical, biological and storms that contribute to characterize it almost uniquely.

The soil has a number of macroscopically observable differences such as color, texture, grit and structure that make it easily identifiable. The change of one of these components, in conjunction with certain chemical modifications related to the decomposition of buried organic elements, is visible at the time when the stratigraphic excavation it starts.

Stratigraphic excavation is the archaeological technique developed to collect and document data, available at a given site, about the human activities that have taken place there and the environment with which they interact.

Table 9.1 General guidelines to follow as part of an excavation

Define the jurisdiction	1. Who oversees the case?
	2. If the remains should result as archaeological and not forensic, who is responsible?
Assessment of the case	Create an action plan, a list of experienced staff and the list of equipment needed, starting with questions such as:
	1. What is the potential number of burials present?
	2. How many forensic sites have been definitely identified?
	3. How many people or artifacts are involved?
	4. Are the sites are very recent or not?
Excavation plan	1. Goal: to reconstruct the past activities at the crime scene, retrieve and document all pre-excavation evidence
	2. Coordinate the staff, identify roles and tasks, and manage the materials/resources available.
	3. Excavation
	4. Clean and/or repair the materials used.
	5. Check the acquired documentation
	6. Complete the excavation maps
	7. Seal the acquired evidence
	8. Compile a final report
Conclusion of the operations	1. Complete all documentation
	2. Make copies, if necessary, before submitting them to the case supervisor

The method was developed from the concept of stratigraphy identified in geology, by which the rocks are deposited in layers, with the oldest at the bottom and the more recent ones covering on top. Similarly, the soil layers that were gradually deposited at a site, allowed for the identification of the chronological sequence of the artifacts that were unearthed. Every human action or any natural event, has left a trace in a site that is overlaid on the existing situation and constitutes a "stratigraphic unit" (SU). The stratigraphic units present "physical" relationships with each other: ditches "cut" the terrain in which they are dug, a wall "fills" its foundation pit, an accumulation of garbage "covers" the paving of the road and "rests on" "the wall against which it was thrown, etc. Conversely, the ground "is cut" by a ditch, the foundation pit is "filled", the road paving "is covered" by garbage and finally, with respect to the wall, the garbage "leans against it." Each of these physical relationships indicates a chronological interaction: if the ditch cuts the ground it means that the action of digging the ditch can only occur after the event of depositing the ground soil where it is dug. The Harris *matrix* (or matrix or stratigraphic diagram) is a graphical representation of the stratigraphic relationships between SU; this system, invented in 1973 in Winchester by archaeologist Edward C. Harris, is therefore a visual and immediate synthesis of what was found (Fig. 9.3).

Schematically, we could identify three fundamental moments in an archaeological stratification:

1. Natural depositions pre-existing human action;
2. Stratification resulting from manufacturing processes, disruptive or anthropogenic transformations, corresponding to the lifetime of a site. Within the active man-made phase can also be found natural deposits related to i) large-scale (such as floods or other) or ii) less severe, resulting from the action of natural agents especially meteorological and climate (the slow degradation of wooden structures, of walls, etc.) or bio-genetic phenomena.
3. Depositions arising from degradation after abandonment at the end of human activity on a site, where once again the natural agents take over.

From this simple scheme, we can therefore, understand how there is a constant interactive continuum between man and nature. In this process we can then identify the agents, actions (or processes) and materials (that are being transformed). The agents can be nature or man, the actions (or processes) may also be natural or man-made, the result of actions or processes is found in the materials (that form the sedimentary deposit both natural and anthropogenic) that were originally however at some point were natural, and that can undergo progressive transformations such as to make them totally anthropogenic, that is artificial.

Agents through the actions (or processes) both natural and anthropogenic always lead to a reprocessing, either by displacement of materials or through a transformation: a reprocessing by dislocation necessarily involves transport, i.e. removing the material from one place to take it to a different location (human action), or the material is transported from one place to another (water or wind action), or moves from one position to another (especially by gravity). Transformation refers instead to a modification of the materials (caused by natural or anthropogenic agents) without dislocation, i.e. without external transport.

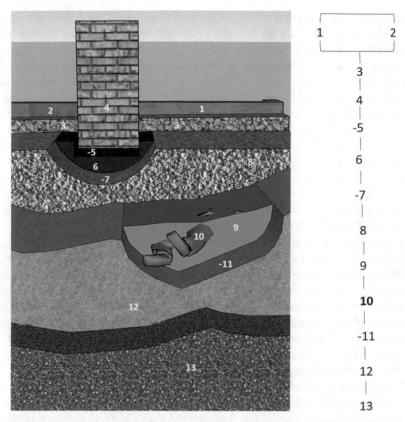

Fig. 9.3 3D expanded section of a forensic archaeological stratigraphy and relative Harris matrix. Notice how the cuts made in the soil are indicated by the '-' sign, while the SU 10 is in bold because it indicates a buried corpse

Any dislocation assumes an original negative evidence (formed by the absence of the material deposited previously, the trace of the removed material) and, after transportation, positive evidence (filler material). The transformation instead produces" neutral" evidence, in the sense that a reprocessing of materials occurs without any dislocation of materials.

In summary, it can be said that every contribution assumes a removal and every removal assumes a contribution, while the changes are the result of reprocesses without transport of materials (therefore without assumption of removal). Schematically, this is the basis of the formation of a stratigraphic deposit either natural or man-made.

There is a further method of excavation besides the stratigraphic type, i.e. nonstratigraphic or arbitrary type. Arbitrary excavation consists of a summary and indiscriminate removal of the soil, performed in order to uncover structures or retrieve objects. The archaeological stratification is not an element of interest for this type of procedure: it is therefore destroyed to reach the target without any prior interpretation. Among the arbitrary excavation categories there is also the practice of so-called excavation "by levels", which is performed by removing horizontal and parallel soil portions of a predetermined thickness, without considering the physical reality of the various components of the layers and their reciprocal relations. This type of excavation begins with the abstract application of the law of superposition and with the assumption that a finding that it is at a lower level with respect to another must be necessarily older, according to a logical hypothesis to be used with great caution in the geological context and which very often can be wrong and misleading in the case of archaeological stratification, which includes, as seen, a dense series of actions, good and bad, of prevailing artificial origin. It is understandable, therefore, how excavation "by levels" has found greater application in the investigation of stratifications of prevalently natural origin for which the removal of deposit portions by successive cuts may allow useful information to be obtained about the formation of the deposit process itself and its duration.

Stratigraphic excavation as mentioned, however, does not remove the soil, but investigates it; the individual elements that make up the layers are first identified and then removed following the reverse order of their deposition: the breakdown of the stratification, therefore, implies its analytic distinction, which is a prerequisite for a synthetic reconstruction of what it has witnessed in the light of the spatial, temporal and cultural relations that linking the different components of the deposit. In the practice of arbitrary excavation, the size of the cut to be made in the ground, its shape and its very location is dictated by the need to achieve maximum results with the least human and economic effort.

Therefore, the chronological reconstruction of the events through the interpretation of stratigraphic data assumes increasing importance. This information provides an indirect or relative dating, but it allows for a an objective consequentiality of events brought to light by the forensic archaeologist.

The great advantage of stratigraphic excavation, in fact, as well as its accuracy and attention paid to the context, is represented by the chronological inferences that can be drawn from the study of the stratigraphic units, as opposed, as seen in the previous chapter, to arbitrary excavation, which is indiscriminate and harmful (Fig. 9.4).

In the relative dating of the excavation, archaeologists must follow two general principles known as *terminus post quem* and *terminus ante quem*. The first, unsupervised *terminus post quem,* indicates a date for the layer that is definitely later than the period under examination. Conversely, *terminus ante quem* it indicates a date for the layer that definitely precedes the period under examination.

To emphasize the importance of this dating system, there is a significant example reported by Matteo Borrini in his book 'L'Archeologia Forense [Forensic Archaeology]' published in 2007. When he was a consultant to the Public Prosecutor of La Spezia "some bones were found, identified as human, in a clearing near a town, together with clearly modern material (cigarette butts, a keychain, a battery of the 1930s–1940s); while intervening with stratigraphic techniques to recover the remains, it was possible to observe the soil sedimentation processes and infer its pedogenesis that suggested the ancient nature of the remains. Moreover, the effects of micro-bradyseism were identified along with evidence of twentieth century

Fig. 9.4 Example of the damage that an arbitrary and unsupervised excavation can cause

building construction which provided an explanation for the appearance on the surface of the bones and their mingling with the objects of the past half century. In the same context, if an arbitrary recovery had been carried out with the concept of predetermined levels, the remains would have been interpreted concealed victims the war period, with clear legal implications, as well as to an error in chronological assessment".

Once the exact contour of the area of interest has been defined, the area adjacent to it changes status: from a search area, it is upgraded to an actual crime scene. The forensic archaeologist, in accordance with the Police forces, will carry out a photographic documentation and a surface survey of the investigated area. These operations serve to contextualize the concealment scene, as well as to highlighting possible routes of access and exit that the offender may have taken, thus allowing for any clue traces to be detected and recorded as relevant case evidence.

The excavation area will be cordoned off to a minimum distance of 4 meters around the area of interest and access will be allowed only to personnel involved in the excavation and recovery operations. This will reduce the possibility of contamination of the scene and at the same time will assure the archaeologist enough space to perform his tasks properly and while maintain the necessary level of cleanliness around the burial area.

After completing these preparations, operations will proceed to remove the filling material in the manner and with the instruments discussed previously.

Keep in mind, finally, that every human action is selective, whether it is of a voluntary or involuntary nature, for this reason the moment of the excavation, and even more so, the documentation phase, are the occasions of maximum likelihood of loss of information, regardless of the extent of our data recovery apparatus (Fig. 9.5). In this sense, there must be a clear strategy and a particular care in the field, in order to achieve subsequent maximum expansion of the data in our possession, through their elaboration and the formulation of hypothetical reconstructions, so as to be able to transform the sediment into a narrative.

The origin and context are extremely important, without which no evidence can be properly inserted into the site puzzle to arrive at a clear and complete representation of the act of deposition, the events surrounding it, and the identification of the

THE CRIME SITE ANALYSIS

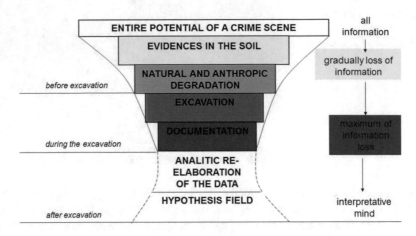

Fig. 9.5 A representation of the loss of information as a result of an archaeological dig

persons involved. Once all the evidence from a site has been recovered and dealt with according to the proper chain of command, a site is usually covered (filled with soil that was originally excavated by the investigators). However, despite the refilling, a scar in the form of a depression in the ground is often visible later in time at the excavation site, however slight. This is because not all the removed ground material is replaced, nor is it replaced with the same orientation or compactness that originally characterized it.

Summarizing, therefore, the recovery of evidence tat is buried or dispersed throughout the surface is normally divided into five major phases:

1. Establish a datum and define a reference grid. The datum is a fixed reference point near the scene (such as a large tree or the corner of a building) which can be found, if necessary, in the course of subsequent investigations. The sub-datum is instead a reference pole placed near the remains at a known distance from the zero point (datum). The lines that run east-west or north-south through the sub-datum are known as base lines. A reference grid crossing the investigated site is extrapolated from these lines. This grid is used to organize all the subsequent collection and excavation activities. It is divided into square units generally of one metre and numbered sequentially. Although not all units are necessarily included in the scene, to every point of the landscape can be assigned a precise reference by adequately measuring its distance to the datum and to the sub-datum;

2. Illustrate the area of investigation. Using rakes and trowels, all debris (leaves, sticks, garbage) must be removed from the surface in order to recover scattered evidence and define the exact boundaries of each element. The top layer of soil (turf grass) is removed with the debris during this phase. Some evidence traces (especially small or dispersed objects) may be collected so as not to forget them it in the next step;

3. Dig up the remains. Each portion of soil that covers the remains of forensic interest is removed according to the procedures set forth above. The bones and artifacts are usually left in place until complete uncovering has occurred.
4. Collect the remains. The bones and artifacts are collected and wrapped in an appropriate way according to their type and nature. The information about the origin is recorded directly on the envelope and on an excavation log, listing all the evidence items.
5. Final cleaning. After removing the evidence, the ground below the removed remains is scraped with trowels and cutting blades, in order to identify any additional evidence. The metal detector can be employed to more quickly locate any underground metal objects.

Obviously, the specific characteristics of a case may require adjustments to these basic concepts. Experts forensic archaeologists know when to modify their techniques to suit the circumstances.

Case Report
A recent example of forensic archaeology applied for the identification of what remains after a disaster is the terrorist attacks in New York of September 11, 2001. This was one of the first times that forensic archaeologists worked on this type of disaster, and it showed the world the benefits of forensic archaeology in the study of disasters.

The *Forensic Archaeology Recovery* group (FAR) was formed at Brown University in Providence, and its mission was to identify the remains at the disaster site and identify them. First, Richard Gould, the team leader, analyzed the remains of St. Nicholas church in Lower Manhattan, which was destroyed during the attack. FEMA (the Federal Emergency Management Agency) and the New York police were working on the case and did not allow them to access the site because FAR had no legal permit. Members of the group then officially instituted the FAR, obtained legal permits and gained access to the site.

In the spring of 2002, they concentrated their investigations on Barclay Street. They removed the refuse in the area and ensured they documented any possible evidence, and its context. Debris samples were hand-dug, sifted, put in sealed bags and labelled as evidence. Also, considerable photographic documentation was produced and 85 % of the site was excavated (Fig. 9.6).

The results were disappointing because the bones found could not be identified as human. Among the refuse were found documents from the Boeing 767 aircraft manufactured in 1991 and some old computers. It could not be ascertained whether the computers had been disposed of before the attack or not, so they could not be used as evidence.

This analysis of a recent disaster site showed that forensic archaeologists should work at disaster sites with the task of quickly responding to certain questions and to finding remains that can be useful as evidence.

During a second investigation in collaboration with the FAR, the New York Fire Department have used GPS to map 41 human bones, this time related to the disaster.

Fig. 9.6 Two images showing *left*, a satellite image of the World Trade Center after the events of September 11, 2001, while on the right, some personal effects of the victims and airplane remains found in various archaeological excavation phases, now in the 9/11 Tribute center

The remains were found in Lower Manhattan and DNA analysis was the main tool used to identify them.

In February 2005, 1,162 of the 2,749 victims of the World Trade Center attack had not yet been identified (as much as 42 %). The news of the identified remains was communicated to the respective families. These families wrote to the FAR and informed them how grateful they were to the archaeologists involved in this type of study.

This event not only demonstrated to the public how forensic archaeologists are needed when there are disasters and crimes, but also that when the work is performed quickly and with the backing of the State, significantly greater results can be achieved.

9.3   Forensic Archaeology and the Integrated Approach

The findings referred to as *Bog Bodies* occurred in various northern European countries (Great Britain, Ireland, northern Germany, the Netherlands, Denmark and southern Sweden). The first specimen discovered has been dated to about 5,500 years ago. The most recent is from the sixteenth century AD. The organisms that live in the marshes contributed to the preservation of the skin and internal organs because of the unusual ambient conditions. Under certain conditions, the acidity of the water, the cold and the lack of oxygen combine and cause the skin to become dark: the preservation of the skeleton is very rare in these findings, because the acid present in the peat normally dissolves the bone calcium carbonate. Many discovered bodies showed signs of violent death; they had been stabbed, bludgeoned, hanged or strangled, and in more than one case the various forms of violence had been perpetrated together. The bodies were often beheaded then deliberately buried in the swamp, using poles to make them sink. Despite the forensic anthropological examinations undertaken over the years, employing increasingly precise and accurate techniques, it is still unclear whether they had been killed and buried in the swamp as a punishment for committing a crime or whether they were sacrificial victims[1] (Fig. 9.7).

When human remains are recovered, the forensic archaeologist must be able to provide the investigators with generic identification of the unearthed remains by defining the subject's sex, age and usually the relevant ethnic group. For this purpose, anthropology is called upon to assist the forensic geo-archaeologists.

Until recently, Forensic Anthropology was considered by the American Academy of Forensic Sciences as the mother discipline that controlled a series of subdisciplines including Forensic Archaeology. This view was resolutely denied and even reversed in Europe, where each of these two disciplines was exercised in its own right, with a profitable collaborative approach. Recently this year, the American Academy of Forensic Science has revised its position converging towards the more correct and profitable European vision. If a given archaeological excavation can

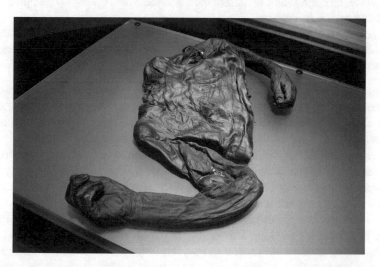

Fig. 9.7 Example of the remains of a mummified body inside a peat bog (so-called Bog Bodies) found in Ireland

---

[1] To view one of the first autopsies performed on one of these mummified human remains of course you can watch the documentary "The Body in the Bog" (1985): http://www.youtube.com/watch?v=aQwB9mdLzlc

begin and end without the necessary contribution of the anthropologist, an anthropological survey cannot exist without a archaeological excavation.[2]

The support afforded by forensic anthropology starts when the forensic archaeologist identifies and recovers the remains of a decaying or already decomposed corpse. Every good archaeologist should be aware of the basics of anthropology in order to be able to properly analyze, retrieve and document the discovered remains. Following a proper archaeological excavation and analysis of the remains at the crime scene, the anthropometric and taphonomic measurements of the remains must be analyzed in order to establish the original position of the victim. Subsequently the analyzes proceed in the laboratory, where, thanks to the study of the hair material, but also of the insects, plants and fingerprints, the experts will attempt to ascertain how much time has passed since the subject's death. Accordingly, the next step is the execution of silicone rubber casts from which positive reliefs are recovered made of resin or ceramic material which will serve both for the museum exhibits and to give the possibility to a greater number of researchers to simultaneously analyze the same subject (Fig. 9.8).

Let's say that the intervention in the laboratory is the most important phase of the study, in fact, it is precisely in the laboratory that the anthropologist works directly with the bones of the subject, first cleaning them, then reassembling the material found and finally a deeper study trying to reconstruct the life of the remains. In the laboratory, as mentioned, the anthropologist focuses on the study of hair and the hair material, insects, plants and fingerprints thus trying to determine the PMI (postmortem interval, i.e. the time interval elapsed since death) and only after this it may be possible, through facial reconstruction and the overlapped photographs, to give provide a face for the subject.

To achieve real and effective results, the forensic anthropologist must have considerable knowledge of the human anatomy, so much so that there are fundamental parameters that he must necessarily be able to identify:

1. *Ethnicity*: easily identified by the shape of the skull and long bones whose parameters to be taken into consideration are, respectively, the shape of the nose, the shape of the orbits, the dental arch, the forward protrusion of the mandible relative to the front as well as the femur-tibia and humerus-ulna ratios;
2. *Gender*: identification of the sex which can be carried out based on the structural difference between male and female (an example may be the diversity of the shape of the pelvis or the difference in diameter of the head of the humerus and the femur, or even the length of the root of the canine teeth);
3. *Age and stature*: respectively identified by observation of teeth, pelvis and skull, and by measuring the long bones;
4. *Presence or absence of trauma*: in fact, it is important to establish whether there were injuries before or after death, and from which type weapon they had been caused, and it is also important to see if there are old fractures which have since recomposed or if medical devices or surgical implants are present.

---

[2] See the Proceedings of the International Conference of the Society of American Archaeology (SAA 2013) – Forensic Archaeology Section.

Fig. 9.8 A moment when the bones are found in a burial

In any case, the role of the forensic anthropologist has been fundamental through-out the centuries; in fact, consider that in the context of forensic medicine there are numerous cases of famous identifications made thanks to teeth and bones: in 1477, at the end of battle of Nancy on May 5, Charles the Bold, Duke of Burgundy was recognized also from his upper teeth.

Another example is that of Prince Napoleon IV who died in South Africa in 1879 as a result of an ambush by the Zulus, when the corpse was returned home in an advanced state of decomposition, it was identified by Dr. Evans by means of the gold fillings he had fitted himself.

Finally, two examples closer to us: the discovery of charred remains found in the Reich chancellery and attributed to the Führer according to the prosthesis and the inlays visible in an x-ray of the skull of Adolf Hitler carried out on the dictator after the 1944 assassination attempt, when he had been seriously injured; the remains found in 2012 in a car park in Leicester, in the UK, were attributed to King Richard III of England. Experts from the University of Leicester had initially diagnosed that the remains belonged to a man in of approximately 30 years, whose skeleton revealed a dozen injuries (eight to the skull, which are potentially lethal), spine curved from scoliosis (all characteristics related to the sovereign described by Shakespeare as a hunchback, who died at 32 years in the battle against the troops of Henry Tudor, later to become King Henry VII of England). Then an accurate carbon 14 dating was performed on the ribs, revealing that the skeleton belonged to a man who died between 1455 and 1540 (Richard III died in 1485). Finally, to close the evidentiary framework, a DNA was performed on the remains and a comparison was made with those of a Canadian furniture maker based in London, Michael Ibsen, direct descendant of Anne of York, sister of Richard III.

In Europe, and particularly in France, we have the first of Forensic Anthropology studies from the late eighteenth century.

In 1846 in the cemetery of Église Sainte Marguerite of Paris, was found a skeleton that Dr. Millicent recognized as belonging to the Dauphin of France, Louis XVII, son of Louis XVI and Marie Antoinette, who died in the prison of the Temple in Paris, at the age of 10 years. After a careful study of the jaws and teeth, other colleagues, the doctors Recamier and Backer, hypothesized instead that the corpse was that of a boy aged 14–16 years and not the age of the Dauphin.

The American Dwight Thomas (1843–1911) in 1878 published the first guide on the use of human remains in forensic practice, along with George Dorsey (1869–1931) and HH Wilder (1864–1928), who had simultaneously also published the manual for the identification of persons living or dead.

Much of what is known about the decomposition of the human body comes from one place: the Body Farm, a research facility covering approximately three acres of land near medical center of the University of Tennessee in Knoxville. Founded as the Anthropology Research Facility in 1980, it became famous in 1994 thanks to the novel by Patricia Cornwell that features detective Kay Scarpetta.

It was founded thanks to the efforts of anthropologist William Bass, university professor, now retired, author with journalist Jon Jefferson, of the enthralling publication The Real Body Factory, that in 20 chapters covers the history of this, to say the least, "original" institution. In this place have "rested" hundreds of bodies, and the observation of their remains has helped to solve numerous crimes and mysteries.

Forensic Anthropology must be grateful to Bass and his assistants for the knowledge and the results obtained over decades spent in the study of thanatological processes of the human body.

From the beginning of his career, W. Bass behaved like a detective always ready to rush to the crime scene, not limiting itself to examining the bones of corpses on the table in a laboratory or under a microscope.

He was assigned his first forensic anthropology case when he was teaching at the University of Kansas in Lawrence, he was given the remains of a woman, collected in a cardboard box, which required identification.

In particular, he was requested to establish the cause of death from signs retained on the bones, distinguishing those lesions attributable to ante-mortem, post-mortem and taphonomic origins. Bass identified the victim, but was not able to determine the causes of death and the case remained unsolved and, however, that skeleton accompanied him for many years in the lectures and seminars, educating thousands of students, detectives, forensic pathologists and forensic anthropologists.

These were only the beginnings of Bass's anthropology career, but the real breakthrough came in the late 1960s, with the call to the University of Tennessee, in Knoxville. At Knoxville, his academic progress blossomed and brought the direction of a National Anthropology course.

It all started there, and thanks to a glaring error by Bass himself, the first seeds were sown for the establishment of the Body Farm.

At the end of 1977 the anthropologist was called to identify the remains of a corpse without a head and in an advanced state of decomposition, probably a victim of murder, protruding from a tomb in a small family cemetery south of Nashville. The police suspected that the murderer had thought of hiding the corpse hastily in the last place where he would be sought: a 1864 grave.

Bass quantified the time passed since the death as between 2 and 6 months. In reality the corpse belonged to the legitimate occupant of that grave, Colonel William Shy, who died in the Civil War, whose body had been dragged out of the cast iron coffin by a defiler of tombs.

"The spectacular error" caused a sensation, and the news spread around the world, highlighting how anthropologists, forensic scientists and pathologists, did not have a sufficient understanding of the post-mortem processes. It was thus that Bass became a pioneer in research activity aimed at the observation of the decomposition of corpses, by mapping the passage of time and documenting every possible variable. To do this, about an acre of unused land was made available, behind the medical center of the University of Tennessee. There he obtained his first body, the first body donated to the Department of Anthropology and there was located the building, still hidden by a high wooden fence that William Bass then referred to as "the scientific establishment of which I am most proud": the Body Farm.

Thanks also to someone like Bass, today forensic anthropological analysis is characterized by a precise and accurate process, adaptable to any environment.

Although most adult skeletons have the same number of bones (206), there are no two identical skeletons. Therefore, the observation of individual specific skeletal traits frequently leads to the identification.

From the shape of the skull it is possible to distinguish three major human ethnic groups: the Caucasoid, Negroid and Mongoloid. The main parameters taken into consideration are the shape of the nose; the inferior nasal edge; the shape of the orbits; the width of the cheekbones; the dental arch; the shape of the teeth; prognathism; the femur-tibia ratio; the humerus-ulna ratio; the anterior femoral curvature.

Caucasian individuals (caucasoid subjects) have the narrowest face with a high nose and prominent chin. Individuals of African origin (negroid subjects) have the wider and lower skull with broader nasal openings. People of Asian origin (mongoloid subjects) have forward projected cheekbones and particular dental characteristics.

Frequently hair is found still attached to the skull. In this case, by examining the color and the microscopic structure it can be, even easier to define the ethnicity.

Sex determination can be made based on sexual dimorphism (that is, the structural differences between male and female). There are, in fact, some anatomical features that differ anatomically according to sex. The main one is the shape of the pelvis.

Some female characteristics are: slighter bones; greater depth; wider pelvic bone upper opening; wider and deeper pelvic cavity; shorter wider sacrum with with less curved upper part; smaller obturator foramina with oval-triangular shape (the male ones are round); wider lower pelvic opening and more mobile coccyx; ischial spines less pronounced towards the inside of the pelvic cavity; smaller acetabula and more frontally directed; shallower symphysis pubis; wider and more rounded pubic arch (forms an angle of 90° with respect to the male of 60°).

These are only a few differences, but they are enough to show that for an expert is not difficult to determine the gender of the body under examination. Other structures which may be considered are the diameter of the humeral head and femur, the length of the roots of the canines, the linea temporalis, the preauricular sulcus, etc.

The age can be estimated by the observation of some elements such as the dental eruption, the root of the third molar, quantity and condition of the dental enamel, the symphysis pubis, sacroiliac joints, cranial sutures, periauricular margins, epiphyseal welds, any arthritic changes in the spine and sternal ribs.

Height may be estimated with the measurement of one or more long bones, preferably the femur or tibia. This measurement of the maximum length of the bone, obtained with osteometric tables (a kind of measuring tape with sliding extremity) can be inserted into a mathematical formula based on ethnicity and sex. If the bones are not complete, however, the estimate is not so reliable. The diagnosis of age, in fact, is of low accuracy in adults if only the cephalic is present and there the best accuracy is a in the range of about 10 years. In this case, the absence of the remaining bone (including the most important anatomical districts for determining the age such as the pubic symphysis and the ribs) will require microscopic studies on the tooth to obtain a more precise diagnosis of age.

Having taken a histologic section by grinding on a Lapping Machine and inclusion of Canadian balsam, we proceed to its analysis under an optical microscope in order to assess and quantify the following parameters according to the method of Gustafrason-Johanson: enamel abrasion, secondary dentin deposition, degree of periodontal disease, transparency of root dentin, secondary cement layers, degree of periodontal disease, root resorption.

It must be established if trauma occurred before or after death (from the state of tissue reaction), what type of weapons were involved (firearms, knives etc.), how many shots were inflicted and so on.

It is also important to assess if there are old now welded fractures, presence of medical/surgical devices (screws, plates …), congenital deformations. All to narrow the hypothesis field and facilitate identification.

The analysis of the marks left by saws on human dead bodies (e.g. in cases of depreciation) was initiated in the US thanks to the anthropologist Steve Symes of Memphis. While odontological identification is useful for the identification of living subjects, for example in the event of aggression in which the victim presents bite marks. In such cases, the comparison between the lesion and the teeth of a suspect attacker can lead to his exclusion or positive identification.

To determine the age of mature adults, microscopic methods are used. If, in fact, from the appearance of the joints it can be understood if the skeleton belongs to a young adult, to have a good estimate of the person's age at the time of death it is sufficient to note the slight degree of wear of certain joints of the pelvis.

However, for older subjects the most reliable methods are those that study the microscopic structure of teeth or bones that can reveal the structural units, at the cellular level, known as osteomas.

According to their shape and distribution, it is possible to ascertain the species of origin, age and, sometimes, the presence of diseases.

In order to arrive at an identity, the second identification phase consists of the comparison of genetic, dental or physiognomic characteristics of the corpse with those of next of kin:

1. DNA tests aimed at the genetic comparison of the corpse's genetic structure with that of the mother, the father and/or children;
2. dentistry specialized exams, to be used in the comparison of registered dental restorations by the dentist (contained in the medical record of the hypothetical dentist of the deceased) with those which may be visible on the corpse;
3. anthropological examinations to be carried out based on the comparison between the morphology of bone structures visible on radiographs in any life to the dead with the morphology of these structures be studied on his body or we proceed to a physiognomic comparison.

In the case of historical subjects, of course without health documentation, the only viable procedures are those of genetic comparison and physiognomy.

Over the years, numerous attempts have been made, in Europe (mainly in Italian and English laboratories), to extract DNA from the bones and teeth of the skull and amplify it through the most cutting-edge molecular biology techniques such as PCR (Polymerase Chain Reaction).

The extraction of well-preserved genetic material allows, in fact, to compare the genetic constitution of a corpse with that of other relatives alive or dead also, in order to achieve a more reliable judgment of identity. However, the feasibility of this investigation is, often, conditioned by the poor condition of the skeletal remains.

Molecular bioarchaeology comes to the rescue of the archaeologist and helps to reveal the details useful to compose the stages of life of all the days in a given period; it is an innovative line of research that involves the construction of new instrumental diagnostic systems carried out on bone and dental tissue for reconstruction of biological events lost in the arrest of the vital processes.

The investigations are of paleogenetic and paleopathological nature and, in particular, of paleoserology, of paleogenetics of the "discontinuous characteristics". In addition, bio-archeology entails dental anthropology research, physiology and pathology research related to work, trauma, epidemiological situations, etc.

Molecular bio-archeology uses the mitochondrial DNA (mtDNA) to study the similarities between different human remains found in various regions. Of great interest are the studies in molecular Bioarchaeology regarding paleopathology and which are referenced in modern diseases, in other words, the study of hereditary diseases through DNA. In this case, nuclear DNA extracted from the remains available is amplified, with particular techniques, to verify the presence of mutated genes whose sequence is, however, known in specific samples of "ancient DNA".

Within the Department of the Italian State Police, an inter-directorate mission group was established for the effective coordination of initiatives and activities undertaken by the staff of the State Police for the identification of victims of natural disasters, both in Italy and abroad, the DVI (Disaster Victim Identification) - Mission Group.

The task force is made up of different professionals, each with different tasks and profiles: fingerprint expert, forensic biologist, pathologist, psychologist, computer scientist, nurse, the crime scene photographer-video cameraman, foreign language expert. Most of them have already been employed by the Scientific Police Service and worked in the case of the Asian tsunami disaster, for a full year from the date of the event, and in the case of the attack in Sharm El Sheik for the identification of all Italians and contributing to the identification of foreign victims.

The DVI mission group operates according to a specific intervention protocol adopted by the Interpol Assembly in 1997 (guide for the identification of disaster victims) which envisages the following key moments: on sight recognition, personal effects, physical features, examination of the tooth structure, fingerprints, anthropological examination and any other activity aimed at identifying (eg genetic identification). Its services are requested, in the cases mentioned above, by the Office of the Public Security Department, while the Central Anti-Crime Department of the State Police (DAC) is responsible for management of not only operational aspects but also of organization, logistics and management.

Forensic Anthropology can be, therefore, considered as the scientific discipline that studies the skeletons or skeletal remains for judicial purposes in support of a archaeological investigation. For this task, it has its roots in multidisciplinary knowledge bases, which together give the same a uniqueness and scientific independence. Such knowledge comes from medicine, particularly from forensic medicine, anatomy and comparative physiology, biology and genetics, physical anthropology, osteology, archaeology, geology, entomology and botany. Furthermore, from these disciplines it also takes the means for research and judicial investigations.

In this context, a specific mention is deserved by Cesare Lombroso (Verona, 1835 – Turin, 1909), founder of criminal anthropology as an autonomous discipline, who assembled diverse knowledge fields, partly ancient, partly contemporary. This non-original method was his greatest scientific limitation, while allowing him to gather orientations and trends that were circulating in European culture, summarizing them in a new approach that moved the focus from crime to delinquent subject, elevated to a study category in human Sciences. The contradictions that marked all of his work are at least two: the claim that a scientific study could be carried out on the nature of criminal individuals, using, however, a very approximate methodology of investigation; and having proposed an interpretation of social pathologies following on the basis of a reformist and progressive ideology – in which he believed profoundly – but with consequences and applications which included, for the defense of that society of which he declared himself a loyal member, no less repressive and segregating instruments than those in force in the Italy of his time.

Among the hundreds of individuals who constituted the Lombroso study material, of particular importance was the "brigand" Villella Giuseppe Motta Santa Lucia in Calabria, who, through the efforts of Lombroso, became a famous case in the world. He probably saw him only once, already a corpse, in the Pavia hospital room in August of 1864. At the time of the autopsy, this case must not have seemed particularly important to him, so much so that the nurse on duty only kept the skull and neither the portrait nor a cast of the face were executed. But 6 years later, reviewing this finding, occurred the so-called *experimentum crucis* of Criminal Anthropology. Lombroso discovered a remarkable median occipital fossa, which he interpreted as the return of ancestral somatic forms to modern humans, thus explaining the criminal behaviour of the subject. Obviously, the Villella skull – whose autopsy evidenced the terrible conditions of hygiene and nutrition in Italian prisons - became a real scientific relic, carrying within it the hard evidence, sufficient to validate the theory of specific physical traits of criminal individuals. In fact, this case could only reflect the serious inconsistencies in the Lombrosian "experimental method" which consisted of accumulating different and often divergent observations working on a reductionist and dogmatic axis, with data often of an impressionistic, second hand nature, collected at different times, from disparate viewpoints, with non-overlapping criteria and methods, or too small to be significant, united in arbitrary frameworks, with abuse of analogies and lack of appropriate control groups (Fig. 9.9).

The final lesson is that if many of the ideas of Lombroso's time are outdated, others survive in different forms but always "historic" in context. Truth and error coexist in every era, and this finding requires caution (but not scepticism) in reading the results of scientific research, where the emphasis is precisely on the word search.

Fig. 9.9 *Left*, Cesare Lombroso; *right*, the skull of Giuseppe Villella

9.4 Survey Criticalities

The future of a discipline called Forensic Archaeology depends on the commitment of its practitioners to uphold a level of quality, competence and professionalism. In Italy, unfortunately, that professionalism is not frequently encountered and forensic archaeologists are struggling to prove their importance during forensic investigations. On the contrary, outside the Italian borders, Forensic Archaeology is not only a discipline held in high esteem, but it is expected in the official protocols of judicial investigations with great advantage for the success of investigations. For this reason, the possibility should be discussed concerning an increase in the use of forensic archaeologists in Italy through a capillary and vocational training effort and the use of international bodies acting as guides and coordinators in international investigations.

In Italy, a possible solution might be to combine these suggestions in one direction leading to the compilation of a handbook, as shown in Table 9.2.

Table 9.2 Guidelines for a good forensic archaeologist

1.	Ability to understand archaeological problems; skills in studying terrain, digging it and rapidly documenting it
2.	Familiarity with the most recent techniques of investigation during the pre-excavation, excavation and post-excavation phases; understanding the relative advantages and limitations
3.	Basic understanding of the components of the human skeleton and their anthropological significance; familiar with taphonomic terminology
4.	Wide knowledge of criminology
5.	Extensive knowledge of the legal system
6.	Ability to manage a survey as an individual professional and as a team
7.	Ability to provide written and oral documentation of the case to superiors; ability to promptly and precisely synthesise a reconstruction in the courtroom
8.	Understanding when is required the intervention of an additional expert at the crime scene
9.	Willingness to constantly update their skills

Chapter 10
Quality, Expertise, and Ethics in Forensic Geoscience

Pier Matteo Barone

Abstract During geoforensic investigations, a typical ethical code must obviously primarily deal with the behaviour of the geoscientist at the crime scene and in the courtroom in relation to the people and evidences with whom he comes in contact. When reasoning or a demonstrating an argument, one should strive for simplicity and conciseness. The professional forensic geoscientist (just like every expert) should be aware of this and should actively cooperate in obtaining the best possible resolution of the case.

Keywords Forensic geoscience • Forensic expert • Ethics • Methodology • Courtroom

In the thirteenth century in China during the Song Dynasty, appeared the first treatise entitled "Xi Yuan Ji Lu", "Collection of rectified injustices", the use of medicine and entomology in criminal proceedings. A classic example is recorded in that volume:

> [...] A farmer from a Chinese village was killed, slain by a sickle. The use of a sickle, a tool used by farmers to cut rice at harvest time, sustained the hypothesis that the murder had been committed by another farmer from the same village. The local magistrate began the investigation by summoning all the local peasants in the town square. Each of them was asked to bring along their own scythe. Once the peasants had assembled, the magistrate ordered 10 or more of the suspects to put their sickle to the ground in front of them and step back a few meters. It was a hot Sunday afternoon and while the villagers, suspects and magistrates were waiting, a swarm of flies began to buzz around them in the town square. The swarm also begun to focus on one of the sickles that was lying on the ground. In a few minutes they had settled on one sickle in particular. None of the other sickles had aroused the same interest in the flies. The owner of this tool became very nervous, and only a few moments had passed before all the villagers had understood who the murderer. The witnesses of the crime were the flies (the impact of the flies) attracted by the remnants of soft tissues, blood, bones and hair of which had marked the sickle.
>
> The book also offers tips on how to distinguish between a drowning (water in the lungs) and strangulation (broken neck cartilage), along with other elements for the examination of bodies and determination of the type of death (homicide, suicide or accident) [...].

P.M. Barone (✉)
Archaeology and Classics Program, American University of Rome, Rome, Italy

Geoscienze Forensi Italia® – Forensic Geoscience Italy, Rome, Italy
e-mail: p.barone@aur.edu

© Springer International Publishing AG 2017
R.M. Di Maggio, P.M. Barone (eds.), *Geoscientists at Crime Scenes*,
Soil Forensics, DOI 10.1007/978-3-319-58048-7_10

In 1784, in Lancaster, in England there was evidence of the first examples of the application of forensic science in legal practice, demonstrating the increasing use of logic and procedure in criminal investigations.

John Toms was tried and convicted for murdering Edward Culshaw. Culshaw had a gunshot wound to the head. Toms had a flintlock muzzle-loading gun. To load such a weapon, it was necessary to pour gunpowder into the barrel and then press it firmly with a rod and wad of paper. The bullet was then inserted into the barrel and blocked with a second wad of compressed paper, to prevent its escape. Often the heavy bullet would carry with it fragments of the second wad into the wound, as had happened in the case of Culshaw: an investigation of the wound revealed a piece of paper folded several times so as to increase its thickness. Cleaned up and rolled out, it turned out to be a strip of paper torn from a poster or an old newspaper. In John Toms' pocket was found the rest of the poster, its torn side was perfectly matched to the wad of paper. Such "scientific" evidence sent Toms to the gallows.

In 1850 the French doctor and psychoanalyst Bergeret solved a case of infanticide by making use of insects as indicators of the time of death: the body of a child was found walled up in the chimney of a house and Bergeret established that the insect remains associated with the corpse (common flesh flies) indicated a state of decay dating back several years, so the responsibility for the infanticide falls on the the previous occupants of the house, not on those present at the time of discovery.

Until the time of Bernard Spilsbury (1899), forensic science was identified almost entirely with forensic medicine; also, toxicology and ballistics were usually entrusted to the doctor. It was Spilsbury who actively encouraged the creation of other professional sectors such as for example legal physics and chemistry. In a case, with the help of a lab technician, he was able to show that the fluorescence under ultraviolet light of a bandage carried by a murder suspect was radically different from that of a similar bandage strip found at the crime scene; and the accused was acquitted.

In the early 1990s, criminologists began to question the nature of forensic science, with obvious consequences for the ethical and deontological implications. In particular, in 1993 the cause Daubert et al. vs. Merrell Dow proved crucial. The US Federal Court ruled on the criteria for the admission of scientific evidence to the court and assigned the judge a fundamental role in its evaluation. In its ruling it cited Popper's theory of falsification as the foundation of scientific evidence. The "Frye test" of 1929 was replaced by "Daubert decision." The Frye test rule stems from a 1923 decision of the Federal Appeals Court[1] and concerns the "general acceptance" concept: to be admissible in court, scientific evidence must be collected using techniques that have "general acceptance" in the field in which they are applied. The ruling states that: "The exact moment when a scientific principle or discovery crosses the line between experimental status and probative demonstration is difficult to define. Somewhere in this twilight zone the evidential force of the principle must be recognized, and while the court will admit the testimony of experts argued on recognized scientific principles or discoveries, the only thing from which the deduction is taken has to be sufficiently demonstrated having obtained general recognition, in the specific area to which it belongs".

---

[1] Frye v.United States, 293 F. 1013, 1014, D.C. Cir.1923.

The Frye test applies also to the testimony of someone who is considered an expert in a field relevant to the case. If the information is not presented convincingly (citing the scientific literature on the method, the use of the procedure, the limits of the procedure and the general approval of the scientific community), or if the expert's qualifications can be cast indoubt, the test may be declared inadmissible.

With the Daubert decision, the Supreme Court established the principle that judges become the guarantors of scientific evidence. They assess whether the proposed testimony of an expert is both relevant to the specific subject under discussion and reliable, so the opinion presented should be based on the application of scientific method. In this sense, it is important to consider whether it is possible to falsify the underlying assumptions, that must also have been the subject of peer review and publication in professional literature, with knowledge of the intrinsic error probability and if is accepted in the scientific community (Frye test).

Furthermore, the current understanding of the nature of the evidence falls within the concept of forensic science in its unitary sense, recalling the falsificationist epistemology of Popper as the only relevant criterion for defining the science. According to the concept of falsifiability, the task of the scientist is not to convert his ideas into proven truth, but to consider them as hypotheses, as always falsifiable constructs, valid until they are proven wrong. His commitment must therefore aim at falsification, rather than verification and defense of his theories.

In any case, the scientific methodology and the reasoning process seem impotent, if not irrelevant if no account is taken of the management process of a crime scene.

The judge is the official representative of the state with power of *iuris dicere*, to put an end to a dispute, to awarded to one of the contenders his life in the judicial process, to end the days of the life of a citizen in a prison facility. It is, as is known, the exercise of one of the three powers in which is articulated the fundamental structure of the state, which cannot be denied or easily obliterated. The judge is in his own right "an Authority".

The lawyer, however, is not. The observation is as banal as it is obvious, but it cannot be disregarded if patient and diligent analysis is to be applied to the delicate theme of forensicethics. The two roles, that of a judge on the one hand and that of the lawyer on the other, from the institutional point of view, are not on the exact same plane, whatever is said or projected when speaking of the relevant professions as "both operators of the law" and sharing the difficulties facing the "Justice system". The court has power, the lawyer does not.

It is from this obvious consideration that must move the analysis on professional ethics in general and in the specific case of field of forensic geoscience.

Professional Ethics would seem to be a term coined by the philosopher Jeremy Bentham, author in 1834 of an essay entitled "*Ethics, or the science of morality*", to designate a utilitarian doctrine of duties. Over time, the term has been more widely used to attempt to explain the study of certain duties in relation to particular social situations. Today, in common parlance, the term professional ethics means a set of rules (or "standards") useful to properly direct the exercise of a profession, especially if this has implications in the realms of journalism. Essentially, professional ethics is the system of provisions, fruit and expression of moral principles that should inspire the behaviour of those who pursue a professional activity in the service of the common good.

In recent years, numerous codes of ethics have flourished, adopted by national legal systems of individual professions; rightly, then, it has been said that professional ethics is the highest expression of the "living law", because the establishment of its rules is based upon an analysis of reality, through which morality becomes law.

Just as the physician should make every possible attempt to save the patient's life, the absolute value in the field of health sciences, but also at the same time safeguarding the principles of his own conscience, so the forensic geoscientist (as anyone who works in this field) is required to defend their actions, but not renouncing the values of truth and loyalty.

As for the profession of the forensic geoscientist a typical ethical code must obviously primarily deal with the behaviour of the geoscientist at the crime scene in relation to the people with whom he comes in contact. But this is not the only concern. It is necessary to examine the rules governing the "how" the forensic geoscientist must maintain his personal and professional relationships with the judge, but first it appears useful to analyze the legal framework within which are regulated the actions of forensicgeoscientists both with the magistrate and with the defenders of the parties.

So then properly it should instead be examined, and as it were, regulated, every aspect of the personal and professional life of the forensic geoscientist.

Like the one that continues, albeit with increasing moderation, to be held in the areas of professional life of other operators whose conduct also in "private life" comes under scrutiny to verify the existence of the conditions that warrant the continuation of the activity, it is necessary to underline the importance that the operator of the legal professions imprints his own life conduct according to rigour, professionalism, balance, moderation, avoiding excesses, attitudes unnecessarily over the top, inappropriate language, behaviour that may cause discredit to the legal profession and the same category of membership. In short, we need to go beyond the scientific 'norm', to establish that the ethics of the forensic geoscientist is actualised in every aspect of social and professional life. This is required by the dignity of their profession, it is demanded by that respect for the best interests of the Nation which is the basis on which we rely during the execution of a trial.

It is quite evident and obvious that the geoscientist suffers daily in the course of the "temptations" that are such because, yielding to them, he betrays the above-mentioned professionalism. One can resist everything but temptation, as Oscar Wilde used to say: and also, forensicgeoscientists, as we all know, are subjected to thousands of temptations. The temptation to talk, rather than with on their professional judgments, in front of the TV cameras; temptations to appear like some American TV series star, divorced from reality; the temptations to go forever in search of a "consensus"; the temptations to use their results as a vehicle for advertising or propagandistic outbursts.

We started from the fact that the lawyer and the magistrate, from an institutional point of view, are not comparable or even overlapping roles. However, some contact points between the two functions undoubtedly exist. The judge exercises power, one of the three into which is divided the political structure of the state, for the highest good of the legal system; the defender plays a role of guarantor for the protection – as mentioned – of the same superior interests of the Nation.

The main point of contact that the two roles show, does not resolve otherwise, in hindsight, than in the requirement of having to respond to the expectations that come from citizens; the judge on the one hand, the lawyer on the another, are at the service of an interest that has a great importance, that of the orderly development of relations woven between the parties. Both are creators of an extraordinary phenomenon: the translation of the abstract right into concrete justice.

And then we cannot fail to point out that, seen from this point of view, not "institutional" but methodological, the two functions must be considered of equal order; there cannot be one that dominates over the other, otherwise it would be impossible to give a fair and proper response to the needs of the parties. But it is on the level of dignity that the two roles weigh absolutely in the same manner. It was already clearly said that "the dignity of judges and defensive functions cannot be graded and must, therefore, be mutually recognized".

In light of this, the judgment of the expert (forensic geoscientist, in this case) can play a key role in the investigation and the trial debate.

It may be surprising that, usually, the investigators (Inspectors and Commissioners of Police, the NCOs and the officers of the Carabinieri, and so on) do not have scientific expertise. In Italy, the careers of these professionals are usually marked by a very strong legal bias and only in rare cases you come across an "investigator" who can boast technical or scientific skills without belonging to the RIS (Scientific Investigations Department of Carabinieri), the Scientific Police or a similar body.

When it comes to "technical" or "scientific" questions, the only hope, therefore, lies in relying on an "expert." What happens is that both the Prosecution that the Defense shall appoint one or more technical experts. The parties' experts prepare a technical report (and a plan of attack), and these are presented in the courtroom somewhat aggressively (usually with no regard for the technical/scientific truth). At that point, the judge (rightly desperate) appoints one or more non-partisan experts (the court expert o forse the expert witness?) and asks them to determine what is true in the reports of the parties' experts.

Obviously, this way of tackling the problem is evidently not "perfect". However, it is frankly difficult to do better. Any attempt to make the process "quieter" and better organized by leveraging the preparation of the experts or their loyalty to the "truth" is bound to fail miserably because of the known limits of human nature (tendency to yield to corruption, excessive confidence in their own means, etc.). A classic example are the court experts, i.e. those appointed by the court. These experts, unlike the experts of the parties, vow to give evidence in full respect of the truth and under total impartiality with respect to the parties. Despite this, they show the same rate of fallibility as the "mercenaries" working for the parties and have the same (admittedly not very high) consideration of the Judges.

But why such a consideration? Where is the problem? The experts are or are not a resource of the system?

The first point to understand is that the expert should be involved from the beginning. Ideally, from the early investigative stages. This applies both to the prosecution's experts and the investigators (i.e., usually, the RIS officers), and to the experts of the defense.

The presence of prosecutionexperts and investigators is necessary both to ensure that the evidence is collected and examined in the proper way (which is the usual work of the various "crime scene investigation units") and to guide the investigation in the right direction as the discovered clues emerge and are analyzed. The investigations, in fact, are an "evolutionary" ("incremental") process: as new clues are found and analyzed, the investigation must take the direction that these clues suggest. There is no investigation in which first all the clues are collected and then later analyzed. These two activities proceed together.

The presence of defense experts is needed to ensure the suspect's rights. The laws of all civilized countries explicitly provide for the right of the accused to be represented by an expert of his choice while the evidence against him is examined. Of course, this right has limits and therefore the defendant must be represented by a lawyer as soon as possible. It will then be the lawyer who asserts these rights properly and to their full extent.

Maybe it is a little less obvious that the judge should be support by one or more experts from the early stages of the trial, long before the need clearly emerges (it is not always necessary to enter into a contract to ask questions to a regular collaborator, etc.). This is because not always the "flaws" are immediately evident and it should be precisely the expert who points this out.

It is a known fact that the investigative units that perform best are often those in which the collaboration between investigators and traditional "experts" (in this case "technical/scientific investigators") is more consolidated. In these cases, the experts often act as actual "consultants" for the rest of the unit. This is, for example, the case in which the RIS cooperate with the other Carabinieri units to carry out an investigation together.

The same goes for the defense. The "best law firms" are often those that rely routinely on a network of specialists from the very beginning. The lawyer that "arbitrarily selects" from the yellow pages, the expert he needs on the spur of the moment does not show a great professionalism. Certain professional relationships should exist beforehand and should have been developed and tested before they are needed in the courtroom.

Something similar is true also for the Judges. A judge should maintain a constant relationship with his network of technical consultants, to "absorb" the information that is useful and so know to whom an expert analysis should be commissioned when required. As I have already said, it is not always necessary to stipulate a contract to ask some questions to one's usual collaborators.

Of course, a single expert cannot handle any technical/scientific "hassle" that may arise, from the collapse of a reinforced concrete building to the death of a patient during an appendectomy, perhaps on the way intercepting a matter of wiretaps. However, it remains true that any law firm of a certain level should have its trusted technical/scientific adviser and should maintain, through that person, a series of collaborations with external technical firms and laboratories that can provide support on various themes that concern the law firm. This is all the truer the greater is the degree of specialization of the law firm. A paying client has every right to expect the presence of an engineer on the staff of a civil law firm specializing in accident litigation (road, industrial and so on).

In Italy, the subject of "forensic experts" is in total anarchy. Either party can involve almost anyone in the guise of "expert." Judges have a few more limitations but, in fact, they can also do what they want.

This total anarchy is both one of the main misfortune and yet one of the best elements of Italian law.

It is a misfortune because, in this way, it is quite normal that a one party chooses the "wrong" expert and does more harm than good.

However, it is also one of the better features of Italian law because, in this way, lot of unnecessary and unpleasant discussions about the "credibility" of the expert are avoided. In Italy, quite simply, it is assumed that the parties' experts (for the Prosecution and Defense counsel) are totally unreliable and untrustworthy. In practice, we rely on them just to bring out the technical and scientific aspects that are relevant to the trial. This is, frankly, the only sensible attitude of mind that can be sustained in a courtroom.

Maybe a little less obvious is the fact that, in the eyes of the judges, almost always the court experts (i.e. those that the judge himself appointed to settle the scientific and technical disputes between Prosecution and Defense) are considered to be substantially unreliable. Theoretically, these experts should be highly technically trained and totally "impartial". The law even provides that their views should be given due consideration by the judges. However, it is not uncommon that the judges appoint several experts one after the other because he does not consider the results of the previous examinations to be decisive.

This healthy mental attitude of absolute distrust is what really guarantees the respect for the truth and respect for the rights of the parties. At the end of legal proceedings based on skepticism, as is typical of the Italian case, what emerges as the winner is often (but not always) the truth based on the facts. In trials based on the credentials of the experts and lawyers (as often happens in the US) what emerges as the winner is often the economic strength of the party who can afford the most spectacular names.

For this reason, in Italy it is often not really necessary that the expert be a luminary. What is really needed is that he is a serious, honest and collaborative professional who is able to work side by side with the assisted party and with the people in his team. In this sense, his communication skills are as essential as his technical capabilities.

Asking the "right" questions to the expert, then, is an art. On the one hand, for the asked questions to it must be possible to give a sensible answer. On the other hand, we should avoid asking questions whose answer will benefit the other party (which will have the decency to pay for its own expert for these services). In Table 10.1 is a list of advice to be followed to make the most of an expert.

Table 10.1 Guidelines for asking the "right" questions to an expert

1.	All the available documentation must be read
2.	The facts must be carefully reconstructed (people, places, times, events, etc.)
3.	The arguments of the other party must be carefully examined
4.	One's own thesis must be formulated (based on one's reconstruction of the facts)
5.	The points where the theses of the parties diverge must be identified
6.	Questions must be formulated that allow for discernment between the two reconstructions
7.	We must be careful not to ask questions that guide the reconstruction of the facts in the wrong direction (the counterparty will surely provide this important service, necessary for the establishing the truth)

Those set forth above are some, general, suggestions that can help a hearing between the judge, prosecution, defense counsel and experts, searching for what could be the criteria to establish a correct and clean relationship with the trial. But the fundamental standards have always been, and still remain, fair play and correctness.

Loyalty corresponds to that attitude of honesty and moral rectitude, of attachment to one's duty but with absolute respect for one's own and the other's dignity, in maintaining professional obligations towards others.

In fairness, it means the immanent reference to a certain "attitude" that a professional must have, or at least promise himself to hold, in dealing with those who encounter him.

If we commit ourselves to follow these basic guidelines, as if they were the two tracks on which our forensic locomotive runs, expecting that our other colleagues will do the same, we will contribute to improving the quality of professional commitments as forensicexpertgeoscientists (and not only), in that common tension dedicated to the performance of our respective duties.

For example, many suggest that the forensic geoscientist should expand his horizons with courses and master's degrees in human rights, international law, precise protocols for taking evidence, ballistics, pathology, biochemistry, osteology, forensic photography, psychology of death and funerary rites in different cultures - these are just some of the subjects that a geoscientist should investigate if he intends to participate in forensic investigations on a global level.

The goal of such training would be to build an understanding of the role and responsibilities of a geoscientist in the context of multidisciplinary investigation.

The study of modern criminology would enable a geoscientist to better understand the scope of research in which he is operating, even in the global hierarchy of the relevant legal system, and the role of other individuals on site.

One of the most important aspects to be recognized as a member of the forensic investigation team is that the artifacts are a link in a chain of evidence that should not be altered or ignored.

What seems clear from the previous chapters about forensic geoscience is the role held by the experience of the individual professional involved in an investigation. If the forensic geoscientist has worked for at least 3 years (three full years, not three seasons!) in different contexts and in different countries and cultures, certainly he will be better equipped to approach the case and may channel his efforts to the understanding of the crime, without wasting energy in deciding what to do.

More generally, when a science is applied, be it physics or chemistry, but also the earth sciences, classifiable in formulas or rigorous evidence-based methods, incontrovertible facts are established that need to be evaluated in the context of evidence; if strychnine is found in a corpse and the correct method was followed to detect it, the presence of strychnine is proven; one must then just find out who put it there (it may even have been put there by the doctor who took the samples or performed the analysis, by mistake or on purpose, but we can be sure, within the limits of human certainty, that the strychnine is there).

But when the doctor makes a diagnosis, he assembles the known data (but it is possible that as many may escape him), he then gives each datum a certain weight and expresses the opinion that his brain has been able to elaborate. This is because there are no scientific or mathematical formula-based rules that allow for a certain diagnosis, as is demonstrated by the fact that there are no serious computer software for performing diagnoses.

Which means that when an exact science is not applied, the expert's opinion can never be expressed in terms of certainty, but only in terms of probability. It is clear that it makes a big difference to say that a certain particle is a residue of a firearm shot with 99% or 70% probability; but unfortunately, there are those who will simply say that the particle "is compatible with a firearm shot residue". Instead, the scientist would answer "based on current knowledge and if the measured data is correct, given that even machines are fallible, given the possible human error, the approximation error inherent in the methods used and applying statistical formulas elaborated above, I can say that my statement is likely to be true with a 72.5% probability and that a single particle does not prove anything." It is up to the court to decide whether the fact that there is such a possibility that the particle is a firearm shot residue and that this is sufficient to show that the suspect had shot his wife; that the particle, if this is what it is, could have ended up in this place for several reasons, before or after the shooting of the wife, etc.

I used the phrase "given the state of the science" because it is known that not a day passes in which notions that were regarded as consolidated, are discovered to be incorrect or dubious.

The word "compatible" which appears in the conclusions of many expert opinions is a logical error that is used by the expert either to avoid admitting that he was not able to give a definite answer or to avoid writing that the prosecution's statement is totally incorrect. An example: the expert writes that "the knife is compatible with the wound". Well, during the trial it will emerge that because the wound had been inflicted with a single edged knife, then it could well have been the accused's single edged pocket knife that had been used. The problem is that it could have been that one with the same probability that it had been a few million other knives!

Instead, we have the exemplary reference of the English situation in which there is the Forensic Sciences Service,[2] subject to parliamentary control. The FSS collects

---

[2] http://www.forensic.gov.uk/

leading experts in each sector and provides scientific support to over 60 foreign countries. It employs about 2,500 people, 1,600 of which are scientists and many are recognized worldwide authorities in their fields. Its seven laboratories are all accredited according to the quality standards UKAS (NAMAS) and BSI QA 9000. Currently this body is suffering somewhat for the reduction of funds, but it is exemplary the clarity with which it faces its own errors. Next to it the Forensic Science Society[3] annually organizes strict tests for aspiring experts, periodically checks that British experts are up to date and have not made mistakes, and revises appraisals if relevant errors are discovered or if improved scientific methods emerge.

In conclusion, we must stress that the logical proof, in the vast majority of cases remains always the most reliable. Statistics show that a crime is a mundane affair and that it is useless to seek crime thriller solutions and that the rules established by a medieval philosopher, famous for the "Occam's razor" logical rules, still apply:

1. All factors being equal, the simplest explanation is to be preferred;
2. Do not multiply the elements and create more subdivisions than necessary;
3. Do not consider plurality if it is not necessary;
4. It is useless to do with more that which can be done with less. In other words, there is no reason to complicate what is simple.

Rather, when reasoning or a demonstrating an argument, one should strive for simplicity and conciseness. The able forensic geoscientist (just like every expert) should be aware of this and should actively cooperate in obtaining the best possible resolution of the case. And let us always remember what the philosopher William of Occam stated: "Other things being equal, simpler explanations are generally better than more complex ones".

---

[3] http://www.forensic-science-society.org.uk/

References

Alestalo J (1971) Dendrochronological interpretation of geomorphic processes. Fennia 105:1–140
Amadesi E (1977) Fotointerpretazione e aerofotogrammetria. Pitagora Editrice, Bologna
Amanti M, Casagli N, Catani F, D'Orefice M, Motteran G (1996) Guida al censimento dei fenomeni franosi ed alla loro archiviazione, vol 7. 109 pp. Miscellanea del Servizio geologico d'Italia, Roma
Anderson RS, Anderson SP (2010) Geomorphology: the mechanics and chemistry of landscapes. Cambridge University Press, Cambridge. 637 pp
Angeli N, Belli R, Galetto M, Toniutti L (2007) L'occhio elettronico che apre nuovi mondi. sem, un innovativo strumento di ricerca e di divulgazione per il museo e lo Science Center. Natura Alpina 57:13–32
Antoci PR, Petraco N (1993) A technique for comparing soil colors in the forensic laboratory. J Forensic Sci 38:437–441
Arnott RD (2010) Introduction to coastal processes and geomorphology. Cambridge University Press., 442 pp, New York
Baggio P, Bellino L, Carraro F, Fioraso G, Gianotti F, Giardino M (1997) Schede per il rilevamento geologico delle formazioni superficiali. Il Quaternario Ital J Quat Sci 10(2):655–680
Barker P (1981) Tecniche dello scavo archeologico. Longanesi, Milano
Barone PM (2012) Archaeology, geophysics and forensic: united we stand, divided we fall. Proceedings of 31st conference of the National Group of Solid Earth Geophysics (GNGTS 2012), Potenza 20–22 Nov. ISBN 978-88-902101-3-6
Barone PM (2016) Understanding buried anomalies: a practical guide to GPR. LAP Lambert Academic Publishers, ISBN: 978-3-659-93579-4
Barone PM, Di Maggio RM (2016) L'approccio alla scena del crimine tramite la geofisica forense ed i cani da cadavere, Il Penalista, Giuffré Editore. ISSN 2464-9635
Barone PM, Bellomo T, Pettinelli E, Scarpati C (2007) Applications of GPR to archaeology and geology: the example of the regio III in Pompeii (Naples, Italy). IEEE, 4th International Workshop on Advanced Ground Penetrating Radar. doi: 10.1109/AGPR.2007.386526
Barone PM, Ferrara C (2016) Georadar between archaeology and forensic investigations. Archeomatica 7(4)
Barone PM, Ferrara C, Pettinelli E (2012) How could archaeo-geophysics help the garbage disposal? A fortuitous discovery in Central Italy. Antiqua. doi: 10.4081/4271
Barone PM, Ferrara C, Pettinelli E, Fazzari A (2012a) Forensic geophysics: how the gpr technique can help forensic investigations. In: Kars H, van den Eijkel L (eds) Soil in criminal and environmental forensics – proceedings of the 6th European Academy of Forensic Science conference, Series: soil forensics. Springer International Publishing, The Hague. 10.1007/978-3-319-33115-7; 978-3-319-33113-3

© Springer International Publishing AG 2017
R.M. Di Maggio, P.M. Barone (eds.), *Geoscientists at Crime Scenes*,
Soil Forensics, DOI 10.1007/978-3-319-58048-7

Barone PM, Ferrara C, Pettinelli E, Annan AP, Fazzari A, Redman D (2012b) Forensic geophysics: how GPR could help police investigations. 18th European Meeting of Environmental and Engineering Geophysics, NSG

Barone PM, Ferrara C, Pettinelli E, Annan AP, Fazzari A, Redman D (2012c) Forensic geophysics: how GPR could help police investigations. Proceedings of the 18th European meeting of environmental and engineering geophysics of the Near Surface Geoscience Division of EAGE, Paris, 3–5 Sept 2012. doi: 10.3997/2214-4609.20143307

Barone PM, Mattei E, Pettinelli E (2013) Non-invasive archaeological exploration in stratigraphically complex rural settings: an example from Ferento (Viterbo, Italy). Archaeol Anthropol Sci. doi:10.1007/s12520-013-0138-3

Barone PM, Di Maggio R, Ferrara C (2015a) Forensic geo-archaeology in Italy: materials for a standardisation, Int J Archaeol Spec Issue: Archaeol Sci. 3 (1-1):45–56. 10.11648/j.ija.s.2015030101.16

Barone PM, Di Maggio R, Ferrara C (2015b) Not necessarily buried bodies: forensic GPR investigations from criminal to civil justice. In Proceedings of the 8th International Workshop on Advanced Ground Penetrating Radar (IWAGPR), IEEE, Florence. doi: 10.1109/IWAGPR.2015.7292681

Barone PM, Di Maggio RM, Ferrara C (2015c) Forensic geoscience during both criminal/civil investigations and courtroom trials. Abstract in Proceedings of the 6th European Academy of Forensic Science Conference (EAFS 2015), Prague. ISBN 978-80-260-8659-8

Barone PM, Di Maggio R, Ferrara C (2015d) Not necessarily buried bodies: forensic GPR investigations from criminal to civil justice. In: Proceedings of the 8th International Workshop on Advanced Ground Penetrating Radar (IWAGPR). IEEE, Florence 2015, pp 1–4. doi: 10.1109/IWAGPR.2015.7292681

Barone PM, Di Maggio RM, Ferrara C (2016a) Forensic geoscience during CS investigations and courtroom trials without a murder. J Forensic Sci Criminol 4 (3). Annex Publishers. ISSN: 2348-9804

Barone PM, Swanger KJ, Stanley-Price N, Thursfield A (2016b) Finding graves in a cemetery: preliminary forensic GPR investigations in the Non-Catholic Cemetery in Rome (Italy). Measurement 80:53–57. doi:10.1016/j.measurement.2015.11.023

Barone PM, Ferrara C, Di Maggio R, Salvati L (2016c) When the crime scene is the road: forensic geoscience indicators applied to road infrastructure and urban greening. Geosciences 6: 50. doi:10.3390/geosciences6040050

Bartolini C, Peccerillo A (2002) I fattori geologici delle forme del rilievo. Lezioni di geomorfologia strutturale. Pitagora Editrice., 216 pp, Bologna

Bashenina NV, Gellert J, Joly F, Klimaszewski M, Scholz E (1968) Project to the unified key to the detailed geomorphological map of the word. Folia Geograph Ser Geogr Phys 2:1–40. Kraków

Beck A, Philip G, Abdulkarim M, Donoghue D (2007) Evaluation of Corona and Ikonos high resolution satellite imagery for archaeological prospection in Western Syria. Antiquity 81(311):161–175

Behrendt R (2012) Introduction to LiDAR and forestry, part 1: a powerful new 3D tool for resource managers. For Source 14–15

Bellotti P, Calderoni G, Di Rita F, D'Orefice M, D'Amico C, Esu D, Magri D, Preite Martinez M, Tortora P, Valeri P (2011) The Tiber river delta plain (Central Italy): coastal evolution and implications on the ancient Roman settlements. Holocene 21(7):1105–1116

Bellotti P, Calderoni G, Dall'Aglio PL, D'Amico C, Davoli L, Di Bella L, D'Orefice M, Esu D, Ferrari K, Bandini MM, Mercuri AM, Tarragoni C, Torri P (2016) Middle to late Holocene environmental changes in the Garigliano delta plain (Central Italy). Which landscape witnessed the development of the Minturnae roman colony? Holocene. doi:10.1177/0959683616640055

Bennett J (2005) The Cromwell street murders: the detective's story. Sutton Publishing, New York

Bevan BW (1991) The search for graves. Geophysics 56(9):1310–1319

Bevan BW (1998) Geophysical exploration for archaeology: an introduction to geophysical exploration, Midwest archeological center special report no. 1. National Park Service, Lincoln

Bierman PR, Montgomery DR (2013) Key concepts in geomorphology. W.H. Freeman & Company, 500 pp

Bintliff J (2004) A companion to archaeology. Wiley-Blackwell, Oxford

Biondi G, Rickards O (2003) Uomini per caso. Mito, fossili e molecole nella nostra storia evolutiva. Editori Riuniti, Rome

Borrini M (2007) Archeologia Forense. Metodo e tecniche per il recupero dei resti umani: compendio per l'investigazione scientifica, Lo Scarabeo

Bowen DQ (1978) Quaternary geology: a stratigraphic framework for multidisciplinary work. Pergamon Press, Oxford. 221 pp

Bozzi S, Grassi A (2009) Il sopralluogo tecnico sulla scena del delitto. In: Picozzi M, Intini A (eds) Scienze Forensi: Teoria e Prassi dell'Investigazione Scientifica. Utet, Torino, pp 27–44

Brancaccio L, Castiglioni GB, Chiarini E, Cortemiglia G, D'Orefice M, Dramis F, Graciotti R, La Posta E, Lupia Palmieri E, Onorati G, Panizza M, Pannuzi L Papasodaro F, Pellegrini GB (1994) Carta geomorfologica d'Italia – 1:50.000. Guida al rilevamento. Quaderni del Servizio geologico nazionale, ser. 3, 4, 42 pp., Roma

Bronstein DA (1999) Law for the expert witness. CRC Press, Boca Raton

Buell J (1997) Bog bodies. Twenty-first Century Books, New York

Canuti P, Casagli N (1996) Considerazioni sulla valutazione del rischio di frana. Atti del Convegno Fenomeni Franosi e Centri Abitati, Bologna, 27 maggio 1994. CNR-GNDCI Regione Emilia Romagna: 29–130 (CNR-GNDCI Pubbl. n. 846)

Capuano E (2010) Milioni di lingotti di tungsteno dorati al posto dei lingotti d'oro vero, http://old. nexusedizioni.it/index2.php?option=com_content&do_pdf=1&id=1882

Carandini A 1991 Storie dalla Terra. Manuale di scavo archeologico. Einaudi

Carella Prada O, Tancredi DM (2000) Il sopralluogo giudiziario medico legale. Ed. SEU

Carrara A, D'Elia B, Semenza E (1985) Classificazione e nomenclatura dei fenomeni franosi. Geol Appl Idrogeol 20(2):223–243. Bari

Carrara A, Guzzetti F, Cardinali M, Reichenbach P (1999) Use of GIS technology in the prediction and monitoring of landslide hazard. Nat Hazards Spec Issue "Tech Tools Mapp Nat Hazards Risk Impact Developed Environ" 20(2–3):117–135

Carraro F (2012) Geologia del Quaternario. L'evoluzione geologica degli ambienti superficiali. Dario Flaccovio Editore., 393 pp, Palermo

Carver M (2010) Archaeological investigation. Routledge, Abingdon

Casadio M, Elmi C (eds) (2006) Il manuale del Geologo. Pitagora Editrice., 808 pp, Bologna

Cassinis G, Solaini L (1946) Note di fotogrammetria. Libreria Editrice Politecnica Cesare Tambutini, Milano

Castiglioni GB (1989) Geomorfologia, Seconda edn. UTET., 436 pp, Torino

Catella G, Mauri P, Raccanelli S, Sommaruga G (n.d.) A methodological approach to evaluate the contamination of groundwater underlayng basins in which industrial wastes where discharged. Proceedings of the 1st international conference on the impact of industry on groundwater resources, 22–24 May, 1996, pp 541–554

Ciavatta C, Cavani L, Gioacchini P, Giovannini C, Montecchio D, Simoni A (2007) L'evoluzione delle tecniche analitiche per la valutazione della qualità agronomica del compost. Fertilitas Agrorum 2(1):65–70

Ciccaci S (2015) Le forme del rilievo. Atlante illustrato di Geomorfologia. Mondadori Università – Sapienza Università di Roma, 576 pp

Claybourn M (ed) (2004) Raman spectroscopy in forensic science. Humana Press, New York

Colman SM, Pierce KL (1991) Summary of quaternary dating methods. In: Rosholt JN (ed) Dating methods applicable to the quaternary. In: Morrison RB (ed) Quaternary nonglacial geology of the conterminous. United States, Geol. Soc. America, DNAG vol. K-2

Colman SM, Pierce KL, Birkeland PW (1987) Suggested terminology for quaternary dating methods. Quat Res 28:314–319

Connor MA (2007) Forensic methods: excavation for the archaeologist and investigator. AltaMira Press, Lanham

Cornwell P (1997) La fabbrica dei corpi. Arnoldo Mondadori Editore

Cosci M, Masella G, Pannuti V (2007) Carta geomorfologica d'Italia – 1.50.000. Guida alla rappresentazione cartografica. Quaderni del Servizio Geologico Nazionale, ser. 3, 10: 177 pp., Roma

Costa JE (1984) Physical geomorphology of debris flow. In: Costa JE, Fleischer PJ (eds) Developments and applications of geomorphology. Springer-Verlag, Berlino, pp 268–317

Cremaschi M, Rodolfi G (eds) (1991) Il suolo – Pedologia nelle scienze della terra e nella valutazione del territorio. La Nuova Italia Scientifica., 427 pp, Roma

Crispino F (2008) Nature and place of crime scene management within forensic sciences. Sci Justice 48:24–28

Crist TA (2001) Bad to the bone?: historical archaeologists in the practice of forensic science. Hist Archaeol 35(1):1–6

Cruden DM (1991) A simple definition of a landslide. Int Assoc Eng Geol Bull 43:27–29

Cruden DM, Varnes DJ (1994) Landslides types and processes. In: Turner AK, Schuster RL (eds) Landslides: investigation and mitigation, Special report 247. Transportation Research Board, US National Research Council, Washington, DC, pp 36–75

D'Orefice M, Graciotti R (2015) Rilevamento geomorfologico e cartografia. Realizzazione – Lettura – interpretazione. Dario Flaccovio Editore, Palermo. 376 pp

D'Orefice M, Graciotti R, Sappa M (2001) L'importanza della geomorfologia negli interventi sul territorio: un esempio applicativo. G.E.AM, Roma, pp 135–138

D'Uffizi M, Sammuri P, Di Maggio RM, Nuccetelli L (2003) Soil laboratory: activities and development. Forensic Sci Int 136(S1-S432):114

Dassié J (1978) Manuel d'archéologie aérienne. Éditions Technip. Paris

Davenport GC (2001) Remote sensing applications in forensic investigations. Hist Archaeol 35(1):87

De Souza T (2009) Ground penetrating radar for forensics. CINDE J 30(4):16–17

Demek J (ed) (1972) Manual of detailed geomorphological mapping. IGU Commission for Geomorphological Mapping. Academia., 344 pp, Prague

Demek J, Embleton C (eds) (1978) Guide to medium-scale geomorphological mapping. IGU Commission Geomorphological Survey and Mapping. E. Schweizerbart'sche Verlagbuchhandlung., 348 pp, Stuttgart

Dennis JN (2008) Evidence submission guideline. Soil samples for forensic analysis. Indianapolis-Marion County Forensic Services Agency

Di Luise E, Magni P, Saravo L (2007) Gli insetti al servizio degli inquirenti. Rassegna dell'Arma

Di Maggio RM (2016) Geological analysis of soil and anthropogenic material. Three case studies. In: Kars H, Van den Eijkel L (eds) Soil in criminal and environmental forensics. Springer, pp 25–43. doi 10.1007/978-3-319-33115-7

Di Maggio RM (n.d.) Caracterización de evidencias y trazas de suelos aplicadas a casos de interés criminalístico en la Policía Forense Italiana. Proceedings of Conference Anàlisis Criminalistico de Suelos, 12–13 April 2012. Instituto Universitario de Investigacion en Ciencias Policiales, Madrid

Di Maggio RM, Barone PM (2016) L'archeologia forense: l'importanza di un corretto scavo sulla scena del crimine, Il Penalista, July 2016. Giuffré Editore. ISSN 2464-9635

Di Maggio RM, Magni P (2016a) Geologia forense. Le indagini sui terreni e sui loro componenti organici ed inorganici. Il Penalista (ISSN 2464-9635), Portali Tematici Giuffrè, A. Giuffrè Editore, Milano

Di Maggio RM, Magni P (2016b) Geologia forense. La falsificazione dei materiali geologici preziosi. Il Penalista (ISSN 2464-9635), Portali Tematici Giuffrè, A. Giuffrè Editore, Milano

Di Maggio RM, Nuccetelli L (2010) Alkaline potassic volcanites in roman soils; silent witnesses in an organized crime case. The 3rd international conference on criminal and environmental soil forensics, Long Beach, Abstracts in webprogramcd, Session 8066, Paper 58551

Di Maggio RM, Nuccetelli L (2013) Analysis of geological trace evidence in a case of criminal damage to graves. In: Pirrie D, Ruffell A, Dawson LA (eds) Environmental and criminal geoforensics, Special publications. Geological Society, London, 384, 75–79

Di Maggio RM, Maio M, Nuccetelli L (2009) Geologia forense. In: Picozzi M, Intini A (eds) Scienze Forensi: Teoria e Prassi dell'Investigazione Scientifica. Utet, Torino, pp 255–266

Di Maggio RM, Barone PM, Pettinelli E, Mattei E, Lauro SE, Banchelli A (2013) Geologia forense. Introduzione alle geoscienze applicate alle indagini giudiziarie. Dario Flaccovio Editore, Palermo

Donnelly LJ, Harrison M (2010a) Development of geoforensic strategy & methodology to search the ground for an unmarked burial or concealed object. Emergency Global Barclay Media Limited, Custom Print, Manchester, pp 30–35

Donnelly LJ, Harrison M (2010b) Geomorphological and geoforensic interpretation of maps, aerial imagery, conditions of diggability and the colour coded RAG prioritisation system in searches for criminal burials, In: 3rd International Workshop on Criminal & Environmental Soil Forensics, Long Beach, California

Dore N, Patruno J (2011) Le nuove frontiere dell'archeologia. dalla fotografia aerea al telerilevamento satellitare SAR, Archeomatica

Dramis F, Bisci C (1998) Cartografia Geomorfologica. Manuale di introduzione al rilevamento ed alla rappresentazione degli aspetti fisici del territorio. Pitagora Editrice., 215 pp, Bologna

Dramis F, Ollier C (2016) Genesi ed evoluzione del rilievo terrestre. Fondamenti di geomorfologia. Pitagora Editrice., 292 pp, Bologna

Dramis F, Guida D, Cestari A (2011) Nature and aims of geomorphological mapping. In: Smith MJ, Paron P, Griffiths JS (eds) Geomorphological mapping: methods and applications, Developments in earth surface processes, vol 15. Elsevier, Amsterdam, pp 39–73

Dupras TL, Schultz JJ, Wheeler SM, Williams LJ (2006) Forensic recovery of human remains: archaeological approaches. CRC Press, Taylor & Francis Group, Boca Raton

Dwyer D (2008) The judicial assessment of expert evidence. Cambridge University Press, Cambridge

Fazzalari E (1974) Il cittadino, l'avvocato e il giudice, Giurisprudenza Italiana

Fenning PJ, Donnely LJ (2004) Geophysics techniques for forensic investigations, in forensic geoscience: principles, techniques and applications. Geol Soc Lond Spec Publ 232:11–20

Ferrara C, Barone PM (2015) Maxwell tra Archeologia ed Investigazioni Forensi. Viaggio nella Scienza V, Ithaca, pp 79–87. e-ISSN: 2282-8079

Ferrara C, Di Tullio V, Barone PM, Mattei E, Lauro SE, Proietti N, Capitani D, Pettinelli E (2013) Comparison of GPR and unilateral NMR for water content measurements in a laboratory scale experiment. Near Surf Geophys. doi: 10.3997/1873-0604.2012051

Fitzpatrick RW, Raven MD (2012) Guidelines for conducting criminal and environmental soil forensic investigations: version 7.0. Centre for Australian Forensic Soil Science (CAFSS), Australia

Franza A (2007) Le nuove tecnologie a supporto dell'attività di indagine nel campo ambientale della P.G., Diritto all'Ambiente

Gale SJ, Hoare PG (2011) Quaternary sediments: petrographic methods for the study of unlithified rocks. Blackburn Press., 325 pp, New Jersey

Gattei S (2008) Introduzione a popper. Laterza, Roma

Ghisalberti A (2000) Introduzione a Ockham. Laterza

Glade T, Anderson MG, Crozier MJ (eds) (2005) Landslide hazard and risk. Wiley, Chichester. 824 pp

Goudie AS (ed) (2004) Encyclopedia of geomorphology. Taylor & Francis Group., 1156 pp, London/New York

Gould RA (2007) Disaster archaeology. University of Utah Press, Salt Lake

Graciotti R, D'Orefice M (2013) L'utilizzo della tecnica della fotointerpretazione in campo forense. In: Di Maggio RM, Barone PM, Pettinelli E, Mattei E, Lauro SE, Banchelli A (eds)

Geologia forense. Introduzione alle geoscienze applicate alle indagini giudiziarie. Dario
    Flaccovio Editore, Palermo, p 320
Gregory KJ, Goudie AS (eds) (2011) The SAGE handbook of geomorphology. SAGE Publications
    Ltd., 648 pp, London
Griffiths P, De Haseth JA (2007) Fourier transform infrared spectrometry. Wiley, Chichester
Grilletto F, Cardesi E, Boano R, Fulchieri E (2004) Il vaso di Pandora. Paleopatologia: un percorso
    tra scienza storie e leggenda. Torino
Groen WJM, Marquez-Grant N, Janaway R (2015) Forensic archaeology: a global perspective.
    Wiley-Blackwell. ISBN: 978-1-118-74598-4
Gualdi E, Russo P (2012) La scena del crimine: Ricerca e recupero dei resti umani, Libreria
    Universitaria Edizioni
Guerini A, Mauri P, Raccanelli S, Sommaruga G (1996) Individuation of a case of pollution (TCE)
    caused by industrial plants dismantled during the seventies. Proceedings of the 1st international
    conference on the impact of industry on groundwater resources, 22–24 May, pp 459–468
Guzzetti F, Carrara A, Cardinali M, Reichenbach P (1999) Landslide hazard evaluation: an aid to a
    sustainable development. Bingampton Symp Geomorphol Geomorphol 31:181–216
Haglund WD (2001) Archaeology and forensic death investigations. Hist Archaeol 35(1):26–34
Haglund WD, Sorg MD (1997) Forensic taphonomy: the postmortem fate of human remains. CRC
    Press, Boca Raton
Harris EC, Brown MR III, Brown GJ (1993) Practices of archaeological stratigraphy. Academic,
    London
Hartlén J, Viberg L (1988) Evaluation of landslide hazard. Proceedings 5th international sympo-
    sium on landslides, Lausanne, 2, 1037–1058
Hughes PD (2010) Geomorphology and quaternary stratigraphy: the roles of morpho-, litho-, and
    allostratigraphy. Geomorphology 123:189–199
Hunter J, Cox M (2006) Forensic archaeology: advances in theory and practice. Routledge, London
Hunter JR, Brickley MB, Bourgeois J, Bouts W, Bourgouignon L, Hubrecht F, De Winne J, Van
    Haaster H, Hakbijl T, De Jong H, Smits L, Van Wijngaarden LH, Luschen M (2001) Forensic
    archaeology, forensic anthropology and human rights in Europe. Sci Justice 41(3):173–178
IUGS/WGL – International Union of Geological Sciences Working Group on Landslides (1995)
    A suggested method for describing the rate of movement of a landslide. Int Assoc Eng Geol
    Bull 52:75–78
Johnson AM (1970) Physical processes in geology. Freeman, Cooper & Co., 577 pp., San Francisco
Johnson AM, Rodine JR (1984) Debris flow. In: Brunsden D, Prior DB (eds) Slope instability.
    Wiley, New York, pp 257–361
Jol HM (2009) Ground penetrating radar: theory and applications. Elsevier Science Ltd, Oxford
Klimaszewski M (1956) The principles of the geomorphological map of Poland. Przegl
    Geögraficzny 28(Suppl):32–40
Klimaszewski M (1963) Landform list and signs used in the detailed geomorphological maps.
    Geogr Stud Polska Acad Nauk 46:139–179. Krakow
Kuptsov AH (1994) Application of Fourier transform Raman spectroscopy in forensic science.
    J Forensic Sci 39:14
Larson DO, Vass AA, Wise M (2011) Advanced scientific methods and procedures in the forensic
    investigation of Clandestine graves. J Contemp Crim Justice 27:149–182
Lombardi G, Fucci P, Gualdi G (1983) La petrografia e le indagini medico-legali criminalistiche.
    La Giustizia penale 88:276–285
Magni P, Di Maggio RM (2013) Scienze Naturali ad Applicazione Forense (Zoologia, Botanica
    e Geoscienze). In: Valli R (ed) Le indagini scientifiche nel procedimento penale. A. Giuffrè
    Editore
Mallegni F (2008) Memorie dal sottosuolo e dintorni. Pisa University Press
Marchetti M (2000) Geomorfologia fluviale. Pitagora Editrice., 247 pp, Bologna
Marella GL (2003) Elementi di antropologia forense. Ed. Cedam

Marumo Y, Seta S, Sugita R (1999) Les éléments de preuve géologiques dans l'enquete judiciaire. Sci Légales, RIPC 474–475:75–84

Megyesi MS, Nawrocki SP, Haskell NH (2005) Using accumulated degree-days to estimate the postmortem interval from decomposed human remains. J Forensic Sci 50(3):618–626

Mellett JS (2011) Clandestine graves: geophysical methods used in their discovery and subsequent exposure. Forensic Magazine

Miall AM (1996) The geology of fluvial deposits. Springer Verlag, Heidelberg. 502 pp

Milsom J (2003) Field geophysics. Wiley, England

Montaldo S, Tappero P (2009) Il Museo di Antropologia Criminale. Cesare Lombroso, UTET, Torino

Morgan RM, Bull PA (2007) Forensic geoscience and crime detection. Identification, interpretation and presentation in forensic geoscience. Minerva Med 127:73–89

Mori E (2012) La drammatica situazione delle scienze forensi in Italia. Earmi.it

Munsell Color (2000) Soil color charts. Revised washable edition by Greta G. Macbeth, New Windsor

Murray RC (2004) Evidence from the Earth. Forensic Geology and Criminal Investigation, Mountain Press Publishing Company, Missoula

Mussat AE, Aftab Khan M (2003) Esplorazione del sottosuolo: una introduzione alla geofisica applicata. Zanichelli, Bologna

Noller JS, Sowers JM, Colman SM, Pierce KL (2000) Introduction to quaternary geochronology. In: Noller JS, Sowers JM, Lettis WR (eds) Quaternary geochronology: methods and applications, American geophysical union reference shelf series, vol 4. American Geophysical Union, Washington, DC, pp 1–10

Obledo MN (2009) Forensic archaeology in criminal and civil cases. Forensic Mag 6:31–34

Panizza M (1987) Geomorphological hazard assessment and the analysis of geomorphological risk. In: V. Gardiner (ed) (1986), International geomorphology. Part I. Wiley, New York, pp 225–229

Panizza M (1988) Geomorfologia applicata. La Nuova Italia Scientifica., 342 pp, Roma

Panizza M (1992) Geomorfologia. Pitagora Editrice., 397 pp, Bologna

Parker R, Ruffell A, Hughes D, Pringle J (2010) Geophysics and the search of freshwater bodies: a review. Sci Justice 50:141–149

Piccarreta F, Ceraudo G (2000) Manuale di Aerofotografia Archeologica. Metodologia, tecniche e applicazioni. Edipuglia, Bari

Pierson TC, Costa JE (1987) A rheologic classification of subaerial sediment-water flows. Geol Soc Am Rev Eng Geol 7:1–12

Pringle JK, Jervis JR (2010) Electrical resistivity survey to search for a recent clandestine burial of a homicide victim. UK Forensic Sci Int 202:e1–e7

Pringle JK, Holland C, Szkornik K, Harrison M (2012a) Establishing forensic search methodologies and geophysical surveying for the detection of clandestine graves in coastal beach environments. Forensic Sci Int 219:e29–e36

Pringle JK, Ruffell A, Jervis JR, Donnelly L, McKinley J, Hansen J, Morgan R, Pirrie D, Harrison M (2012b) The use of geoscience methods for terrestrial forensic searches. Earth-Sci Rev 114(1–2):108–123. doi:10.1016/j.earscirev.2012.05.006

Pye K, Croft DJ (eds) (2004) Forensic geoscience: principles, techniques and applications, Special Publications, 232. Geological Society, London

Reading HG (ed) (1996) Sedimentary environments: processes, facies and stratigraphy. Blackwell Science Ltd., 688 pp, Oxford

Rebman A, David E, Sorg MH (2000) The cadaver dog handbook: forensic training and tactics for the recovery of human remains. CRC Press, Boca Raton

Rees WG (2013) Physical principles of remote sensing. Cambridge University Press, Cambridge

Regione Emilia-Romagna (1980) Guida alla fotointerpretazione e restituzione tematica. Pitagora Editrice, Bologna

Renfrew C, Bahn P (2006) Archeologia. Teoria, metodi, pratica. Zanichelli

Reynolds JM (2011) An introduction to applied and environmental geophysics. Wiley
Reynolds MP, King PSD (1992) The expert witness and his evidence. Blackwell, Oxford
Ricci Lucchi F (1978–1980) Sedimentologia. CLUEB, Bologna, 3 volumi: vol. I, 217 pp.; vol. II,
    210 pp.; vol. III, 504 pp
Ricci Lucchi F (1992) Sedimentografia. Atlante fotografico delle strutture dei sedimenti.
    Zanichelli., 249 pp, Bologna
Ritter DF, Kochel RC, Miller JR (2011) Process geomorphology. Waveland Press., 562 pp, Illinois
Rolandi V, Cavagna S (1996) Procedure e metodi di indagine tradizionali ed avanzati per
    l'identificazione delle gemme, Hoepli
Ruffell A (2004) Burial location using cheap and reliable quantitative probe measurements. Diversity
    in forensic anthropology. Forensic Sci Int 151:207–211. doi:10.1016/j.forsciint.2004.12.036
Ruffell A (2005) Searching for the IRA "Disappeared": ground-penetrating radar investigation of
    a churchyard burial site, Northern Ireland. J Forensic Sci 50:1430–1435
Ruffell A, McKinley J (2008) Geoforensics. Wiley, Chichester. ISBN: 978-0-470-05734-6
Ryàř J, Stemberk J, Wagner P (eds) (2002) Landslides. Swets & Zeitlinger B.V, Lisse. 734 pp
Schultz JJ (2012) The application of ground-penetrating radar for forensic grave detection. In:
    Dirkmaat D (ed) A companion to forensic anthropology, chapter 4. Wiley, Oxford, pp 85–100.
    doi:10.1002/9781118255377.ch4
Schultz JJ, Martin MM (2011) Controlled GPR grave search: comparison of reflection profiles
    between 500 and 250 MHz antennae. Forensic Sci Int 209:64–69
Schultz JJ, Collins ME, Falsetti AB (2006) Sequential monitoring of burials containing large pig
    cadavers using ground-penetrating radar. J Forensic Sci 51:607–616
Scott DD, Connor M (2001) The role and future of archaeology in forensic science. Hist Archaeol
    35(1):101–104
Scott DD, Connor M, Michael RL (2001) Archaeologists as forensic investigators: defining the
    role. Society for Historical Archaeology
Settimo M (2010) Lo scienziato forense. Origine, storia e presupposti teorici di una professione.
    AIASU
Shepherd EJ, Palazzi DS, Leone G, Masae M, Mavica M (2012) La collezione c.d. USAAF
    dell'Aerofototeca Nazionale. Lavori in corso. AAerea 6:13–32
Simons LM (2010) Guerre di fossili, http://www.nationalgeographic.it /scienza/2010/04/12/news/
    fossil_wars-3625/
Smiraglia C (1992) Guida ai ghiacciai e alla glaciologia. Forme, fluttuazioni, ambienti. Zanichelli.,
    240 pp, Bologna
Smith D (1993) Being an effective expert witness. Thames Publishing, London
Smyth F (1984) Sulle tracce dell'assassino: storia dell'investigazione scientifica. Dedalo
Sommaruga G (2013) Terre e rocce da scavo. Dario Flaccovio Editore
Spagna V (2002) Aero-geologia. Pitagora Editrice, Bologna
Spagna V (2013) Geologia delle frane. Riconoscimento, prevenzione, difesa. Dario Flaccovio
    Editore., 200 pp, Palermo
Spanò G, Tedeschi A (2012) Il tecnico forense. Compiti del c.t.u., c.t.p., perito e arbitro. Giuffré
    Editore
Speranza Cavenago (1980) Bignami Moneta, Gemmologia, Hoepli
Stejskal SM (2013) Death, decomposition, and detector dogs: from science to scene. CRC Press,
    Taylor & Francis, Boca Raton
Strahler AN (1984) Geografia fisica. Edizione italiana a cura di Pellegrini G.B., Sauro U., Zanon
    G. Piccin Nuova Libraria, 664 pp., Padova
Sung T, McKnight E (1981) The washing away of wrongs: forensic medicine in thirteenth-century
    China (Science, Medicine, and Technology in East Asia), Center for Chinese Studies, The
    Universi
Thornbury WD (1969) Principles of geomorphology. Wiley., 594 pp, Chichester
Tricart J (1965) Principes et méthodes de la géomorphologie. Masson et Cie., 496 pp, Paris

Tricart J (1972) Normes pour l'établissement de la carte geomorphologique detaillée de la France: (1:20.000, 1:25.000, 1:50.000). Memoires et Documents, année 1971, n.s. 12, Paris, 105 pp

UNI EN 10802 (2013) Wastes – manual sampling and preparation of sample and analysis of eluates

UNI EN 12457-2 (n.d.) Characterisation of waste – leaching – compliance test for leaching of granular waste materials and sludges – part 2: one stage batch test at a liquid to solid ratio of 10 l/kg for materials with particle size below 4 mm (without or with size reduction)

UNI EN 12920 (2009) Characterization of waste – methodology for the determination of the leaching behaviour of waste under specified conditions

UNI EN 14899 (2006) Characterization of waste – sampling of waste materials – framework for the preparation and application of a sampling plan

UNI EN 932-2 (2000) Tests for general properties of aggregates – methods for reducing laboratory samples

Varnes DJ (1978) Slope movements types and processes. In: Schuster RL, Krizeck RJ (eds) "Landslides: analysis and control". Washington Transp Res. Board, Spec. Rep., 176, Nat. Sci. Acad., 11–33

Verstappen H.Th, Van Zuidam RA (1968) ITC system of geomorphological survey. In: ITC textbook of photointerpretation chap. 7, 2, Delft, 1–49

Walker M (2005) Quaternary dating methods. Wiley., 286 pp, Chichester

Wall W (2010) Forensic science in court: the role of the expert witness. Wiley, Oxford

Warnasch SC (2016) Forensic archaeological recovery of a large-scale mass disaster scene: lessons learned from two complex recovery operations at the World Trade Center site. J Forensic Sci 61(3):584–593. doi:10.1111/1556-4029.13025

Watters M, Hanter JR (2004) Geophysics and burials: field experience and software development. In: Forensic geoscience: principles, techniques and applications. Special Publications. Geological Society, London 232

Werner ED, Friedman HP (eds) (2010) Landslides: causes, types and effects, Natural disaster research, prediction and mitigation series. Nova Science Publishers, Inc.., 404 pp, Hauppauge

Wilson DE (1982) Air photo interpretatoin for archaeologists. London

WP/WLI – International Geotechnical Societies' UNESCO Working Party on World Landslide Inventory (1990) A suggested method for reporting a landslide. Int Assoc Eng Geol Bull 41:5–12

WP/WLI – International Geotechnical Societies' UNESCO Working Party for World Landslide Inventory (1993) Multilingual glossary for landslides. The Canadian Geotechnical Society, BiTech Publisher Ltd., Richmond

Websites

http://gia4cs.gia.edu
http://numistoria.altervista.org
http://www.emfa2017.eu
http://www.forensic.gov.uk/
http://www.forensicgeologyinternational.com
http://www.forensic-science-society.org.uk/
http://www.geologiaforense.com
http://www.geoscienzeforensiitalia.com
http://www.igmi.org/prodotti
http://www.moissaniteitalia.com
http://www.pietrepreziose.eu
http://www.terranea.it
http://www.youtube.com/watch?v=aQwB9mdLzlc

http://www.rai.it/dl/RaiTV/programmi/media/ContentItem-30cf2dfe-cebf-4bf9-ae22-d4700651db05.html
http://www.rai.it/dl/RaiTV/programmi/media/ContentItem-37058a87-d6d3-488b-b784--33cd9eebc360.html

Index

© Springer International Publishing AG 2017
R.M. Di Maggio, P.M. Barone (eds.), *Geoscientists at Crime Scenes*,
Soil Forensics, DOI 10.1007/978-3-319-58048-7

Printed in the United States
By Bookmasters